T0358264

Social Knowledge

Historical Materialism
Book Series

VOLUME 207

The titles published in this series are listed at *brill.com/hm*

Social Knowledge

An Essay on the Nature and Limits of Social Science

By

Paul Mattick

BRILL

LEIDEN | BOSTON

Second edition, with a new Preface by the author. The first edition was published in 1986 by Hutchinson, ISBN 9780091654603.

Library of Congress Cataloging-in-Publication Data

Names: Mattick, Paul, 1944- author.
Title: Social knowledge : an essay on the nature and limits of social science / by
    Paul Mattick.
Description: Second edition. | Leiden ; Boston : Brill, [2020] | Series: Historical
    materialism book series, 1570-1522 ; volume 207 | "With a new preface by the
    author"–Verso. | Includes bibliographical references and index.
Identifiers: LCCN 2019043764 (print) | LCCN 2019043765 (ebook) |
    ISBN 9789004414808 (hardback ; alk. paper) | ISBN 9789004414822 (ebook)
Subjects: LCSH: Social sciences–Philosophy.
Classification: LCC H61 .M428 2020 (print) | LCC H61 (ebook) | DDC 300.1–dc23
LC record available at https://lccn.loc.gov/2019043764
LC ebook record available at https://lccn.loc.gov/2019043765

Typeface for the Latin, Greek, and Cyrillic scripts: "Brill". See and download: brill.com/brill-typeface.

ISSN 1570-1522
ISBN 978-90-04-41480-8 (hardback)
ISBN 978-90-04-41482-2 (e-book)

One sees how subjectivism and objectivism, spiritualism and materialism, activity and passivity, lose their antithetical character, and hence their existence as such antitheses, only in a truly social state of affairs; one sees how theoretical antitheses can be resolved only in a practical way, only through people's practical energy, and how their resolution is therefore by no means only a task for knowledge, but a real problem of life, which philosophy has been unable to solve precisely because it has considered it as a purely theoretical problem.

KARL MARX, *Economic and Philosophical Manuscripts* (1844)

∴

Contents

Contents

Preface to the Second Edition

In an otherwise critical review of *Social Knowledge*, Leonidas Bouritsas kindly suggested that 'Mattick's book deserves a central position in the ongoing discussion' of 'the question of objectivity in the social sciences'.[1] Whether it deserved this position or not, the book was in the event entirely ignored, despite the many, mostly positive, reviews it garnered on its publication. So I was surprised and pleased when an anonymous referee supported its republication in the Historical Materialism series. I am particularly happy about this, because I always imagined *Social Knowledge* as a prolegomenon to the study of *Capital* on which I had embarked in the 1980s, and which has itself just been published as a volume in the same series.[2] Written at such a temporal distance from each other, the two books are not parts of one work, but they are complementary, the first providing clarification of the epistemological conditions for Marx's critique of political economy, and the second exploring the method and uses of that critique.

Rereading *Social Knowledge* in preparation for this re-edition I could not help but be struck by the marks left on it by the period and occasion of its writing. An expanded version of my doctoral dissertation, it testifies to its gestation within the academic – particularly philosophical – discussion of the social sciences after 1970. My own absorption in this discussion shows itself in the brevity with which I treat matters (such as W.V.O. Quine's ideas about 'radical translation') with which general readers, even academic ones, could not be expected to be at home, but which were simply part of the conceptual air I breathed at the time.[3] Most obviously, the book focused much of its argument on the ideas of Peter Winch, a philosopher then taken quite seriously and largely ignored today.[4] I would not make much of an argument for a revival of interest in these ideas, if only because I believe I have adequately criticised

---

1 Bouritsas, 'Book Review: *Social Knowledge*' (1988), p. 128.
2 Mattick, *Theory as Critique: Essays on* Capital (2018).
3 I have made only the smallest corrections of this flaw in the present edition, mostly by providing a few references for the reader who wants to follow up on a mention of some writer or idea. I have altered quotations from Marx's writings to follow newer, now standard, translations. New material, other than minor alterations, is enclosed in square brackets.
4 The most recent attempts to revive interest in his ideas known to me are a set of articles in *History of the Human Sciences*, 13, no. 1 (2000), followed up by Flathman, 'Wittgenstein and the Social Sciences: Critical Reflections Concerning Peter Winch's Interpretations and Appropriations of Wittgenstein's Thought' (2000), pp. 1–15.

them. But that is not to say that issues once approached through a consideration of Winch's work are not worth another look.

Winch's use of Ludwig Wittgenstein's conception of the relation between
'language games' and 'forms of life' to criticise the scientific pretensions of
the social sciences were no doubt still resonating in the 1970s thanks to the
questioning of those academic areas born out of the student upheavals of the
later 1960s. At the same time, the logical-empiricist attempt to provide firm
foundations for the natural sciences as uniquely forms of objective knowledge
had finally to be acknowledged as a failure, even by most of its earlier adher-
ents. Carl Hempel's 1950 article 'Problems and Changes in the Empiricist Criterion of Meaning'[5] demonstrated a continuing inability to distinguish science
from non-science as the positivists had hoped to do; Quine's 'Two Dogmas of
Empiricism' of the following year[6] proclaimed the objects theorised by modern physics not different in nature from the gods of Greek mythology; in 1962
Thomas Kuhn's *The Structure of Scientific Revolutions* seemed to many to under-
mine radically the idea that the history of science represented progress towards
an objective truth; as a grand climax to these developments, the 1969 Urbana
(Illinois) symposium on the structure of scientific theories laid to rest the positivist 'Received View' of science, the published proceedings providing (in the
editor's words) 'a graphic display how philosophy of science – circa 1969 – was
attempting to find a new direction in the aftermath of positivism's collapse ...'.[7]

The two chief modes in which the question how that new direction was to be
charted was posed were the so-called rationality debate and the argument over
realism in scientific theory. The two were connected, in that 'rationality' was
taken to involve a quest for truth independent of the conceptual frameworks of
particular cultures or historical periods, and 'realism' meant the claim that the
concepts and laws of the sciences accurately describe the phenomena investigated. The problem of rationality was directly provoked in the 'human sciences',
especially anthropology, by the dismantling of colonialism, which profoundly
disturbed the earlier assumption of European scientific rationality in contrast
to 'native' adherence to 'tradition'. Within philosophy, the collapse of positivism
seemed to spell the untenability of belief in transcultural canons of rationality
and scientific truth, so that 'science' itself might seem just a peculiarity of a

---

5   *Revue Internationale de Philosophie*, 11, pp. 41–63.

6   *Philosophical Review*, 60, pp. 20–43.

7   Suppe (ed.), *The Structure of Scientific Theories* (1977), p. 241. I myself participated in an
    important attempt to understand the nature of scientific information in non-positivist terms:
    see Harris et al., *The Form of Information in Science: Analysis of an Immunology Sublanguage*
    (1989).

particular culture, calling for anthropological investigation rather than philosophical justification. In the refreshing words of Ian Hacking,

> Philosophers long made a mummy of science. When they finally unwrapped the cadaver and saw the remnants of an historical process of becoming and discovering, they created for themselves a crisis of rationality. This happened around 1960.[8]

This was the context in which I argued that the scientific character of Marx's critique of political economy arose from his ability to question the validity of the 'language game' – political economy – fundamental to the 'form of life' – the production and exchange of commodities as a means to the accumulation of capital – central to the society in which he lived. That is, given that Marx's object of study was a human social order, analysis and understanding of that order, as opposed to simple participation in it, required the construction of concepts other than those fundamental to the order itself. This was made possible by Marx's radically historical approach to society, an approach that identified analytically significant differences between social systems. Rendering impossible a general theory of society, this approach made possible a science of social life based precisely on recognition of these differences.[9] In contrast to political economy, which claimed to discover in capitalism features essential to any possible society, Marx's critical approach to the same subject matter identified features both peculiar to modern society and doomed to disappear when people put an end to this social order. Cross-cultural comparison, in the form of historical investigation, made possible the replacement of economic ideology by a scientific understanding of society.

Notice that this formulation stresses neither 'rationality' nor 'realism'. Of course, one could complain that the economists' belief that their science usefully describes the functioning of capitalist society, despite the constant failures of prediction and the well-known weakness of economic analyses, was, and is, irrational. But it is odd to describe people as 'irrational' when they are following the lessons of their teachers and cooperating in research and teaching with

---

8 Hacking, *Representing and Intervening: Introductory Topics in the Philosophy of Natural Science* (1983), p. 1.

9 This emphasis on historical differences is, of course, another manifestation in my text of the period's intellectual preoccupations, visible in Michel Foucault's 'archaeology of knowledge' as well as in Kuhn's focus on scientific revolutions. In Marx's case, the stress on discontinuity (alongside the continuity of historical change) reflected more the experience of major social-historical transformations, particularly in the form of the French Revolution.

their colleagues, following carefully formulated rules, and, where appropriate, processing masses of data. In the same way, it is not clear how it is analytically useful to describe a Catholic priest as 'irrational' for acting on the assumption that the ceremony of the Mass produces release from sin. The concept of rationality is of as little use in understanding the persistence of belief in economic science as criticising the Azande belief in witchcraft as irrational – despite the fact that (in Evans-Pritchard's famous words) 'witches, as the Azande conceive them, clearly do not exist' – would have helped that anthropologist understand that belief. Marx's anthropology of his own culture sought the origins of the persistence of economic beliefs not in features of thought but in the structure of capitalist social relations, to which thinking is largely accommodated. In fact, he showed, what is remarkable is that people are ever able to break free of such beliefs; that they are, he argued, demonstrates not so much an abstract power of reason as capitalism's ongoing transformation towards a new form of social life, a movement that may be hard to recognise in itself. (In the same way, the development of late-medieval society towards capitalism showed itself in the critique of religious thinking and institutions, such as Spinoza's scriptural philology, and the development of natural-scientific modes of explanation.[10])

On the other hand, such an approach might seem to lead most naturally to a relativism about the categories used to analyse social life; why take discontinuous change as progress towards truth? Of course, it needn't be; it is hard to see, whatever Hegel might have maintained, that the displacement of classical polytheism by Christian monotheism in the late Roman Empire represented progress towards truth about the divine. But what right do I have to say that this is because there is no truth about the divine, or at least no truth of the sort that made Newton's account of the heavenly motions superior to Galileo's?

Let us start by insisting that scientific truth – in the natural as well as the social sciences – is a social construction. The logical empiricists were unable to show that scientific objects and the laws relating them were representations of elements given in 'immediate experience', structured according to the invari-

---

10   For a more satisfying, because more complicated, explanation of this sort, see Brian Eas-
     lea's argument 'that to the extent the mechanical philosophy triumphed over natural
     magic [in the seventeenth century] it did so, at least in part, not solely because it was
     an ingenious philosophy ..., but also because it was regarded as an "establishment" philo-
     sophy that upheld religion and the social order against the perceived threat of natural
     magic and "enthusiasts" while at the same time it legitimated and suggested the feasibil-
     ity of mechanical appropriation of the natural world without impugning the miraculous
     nature of Christ's works' (*Witch Hunting, Magic, and the New Philosophy: An Introduction
     to the Debates of the Scientific Revolution, 1450–1750* [1980], p. 197).

able laws of logic, and so independent of the work of culture. This is because there is no experience not shaped by the systems of conceptual constructs – above all those embodied in language, but we must not forget those embodied in tools and disciplined theoretical and experimental practice – collectively developed over time. Given that we are social beings, science can only be a cultural construction, once we abandon such hypotheses as Descartes's idea that God has implanted the basic elements of knowledge in our minds. This has become increasingly obvious as the production of scientific knowledge, since its seventeenth-century beginnings, has become dependent on ever more complex machinery, mathematical and material, itself based on earlier such knowledge, though detailed study of this aspect of scientific work only took off around the time I was writing *Social Knowledge*.[11]

An anthropological approach to scientific practices, Wolf Lepenies asserts, means regarding science 'as a system of beliefs and activities whose structure has no *a priori* privilege in regard to other forms of knowledge'.[12] But such an approach would be incomplete, because the uses to which systems of beliefs and activities can be put, and hence the reasons for their evolution, are germane to the judgement as to the fitness of such systems to those uses and so also to the explanation of such systems. As a social product, science's purposes and the forms it is given to meet those purposes are socially determined. Those purposes historically have been heterogeneous, combining intellectual and ideological interests with technological goals: particle physics, at its origin strongly impelled by scientific preoccupations, was soon overtaken by the wish to build atomic bombs. The satisfaction of the competitive curiosity of physicists had eventually to coincide with the actual ability to obliterate Japanese cities.

Among the purely intellectual goals of scientific activity, such elements as accuracy of prediction and completeness of explanation have coexisted with features like simplicity, aesthetic beauty, and support for beliefs that seem fundamental to the continuance of a given social order. A science like economics, for instance, while often attempting to measure and analyse data accurately, is more strongly regulated by the wish to uphold certain basic understandings of the social order, such as the 'productivity' of privately owned resources, as well as the maintenance of its existing store of concepts. This is because it is concerned with the analysis of contemporary society, and so comes strongly under the pressure of social power relations and subgroup interests; in contrast, a sci-

---

11   An exemplary study is Galison, *Image and Logic* (1997).
12   Lepenies, 'Anthropological Perspectives in the Sociology of Science' (1981), p. 247.

ence like physics, conceiving its objects as independent of human reference, is freer to concentrate on its predictive and explanatory apparatus.[13]

G.E.R. Lloyd's discussion of the development of critical views about magic in Greece around the fourth century BCE, as exemplified by the anonymous text *On the Sacred Disease*, provides a useful example of a distinction between scientific and non-scientific modes of thought independent of the concept of rationality (and without overlooking the inadequacies of the scientific opinions he examines). Central to the scientific point of view in Lloyd's account is the attempt to maintain a thoroughgoing scepticism. Taking as a starting point – once again! – Evans-Pritchard's account of Azande witchcraft, he notes that 'although many Azande suspect individual witchdoctors of being frauds, there is no skepticism about witchdoctorhood in general ...'.[14] In contrast, '[w]hat is important in the attack expressed by the author of *On the Sacred Disease* is that it is directed against *all* the [magical] purifiers, and against *any* idea that the sacred disease [i.e. epilepsy] or any other disease is the result of divine intervention, indeed against any idea that ritual purifications can influence natural phenomena in any way'.[15] In part this reflects the development of a concept of 'nature' (*phusis*) as a set of phenomena not subject to supernatural causes. But it also reflects 'philosophers' concern with the question of the *grounds* for the views and theories they advanced'[16] reflected in their efforts to develop theories of logic and reasoning.

> The question we must turn to in conclusion is ... this: do the radical developments that occur in either the practice or the theory of reasoning in Greek thought imply any shift or transformation in the underlying logic or rationality itself? ... Certainly new modes of argument, some of them quite technical, can be said to have been *invented* ... Yet it is not the case that the logic itself is *modified* by being made explicit, *except insofar as it is made explicit*.[17]

---

13   Though here too disparate social purposes are served: to take two examples, Newton's mechanics was not independent of his religious views, and Quantum Mechanics developed through a highly conflicted dialogue among practitioners both competitive and cooperative (see Beller's fascinating *Quantum Dialogue* [1999]).

14   Lloyd, *Magic, Reason, and Experience: Studies in the Origin and Development of Greek Science* (1979), p. 18. It may be worth noting the similarity to this of modern views about medical doctors.

15   Ibid., p. 18.

16   Ibid., p. 233.

17   Ibid., pp. 123–4.

It is not the peculiar rationality of Greek science that accounts for its difference from magical thought but 'the exceptional exposure, criticism, and rejection of deep-seated beliefs', visible in a 'generalized skepticism about the validity of magical procedures'.[18]

This basis in generalised scepticism is no doubt in part what accounts for the relatively relaxed attitude of the modern natural sciences to the concept of truth, at least from the eighteenth century on, in comparison to the fierceness with which religious orthodoxy was upheld at the same time. Already in 1726 Newton asserted that in 'experimental philosophy we are to look upon propositions collected by general induction from phenomena as accurately or very nearly true, notwithstanding any contrary hypothesis that may be imagined, till such time as other phenomena occur by which this may either be made more accurate, or liable to exceptions'.[19] It is in the same spirit that the chief picture of reality utilized in present-day high energy physics is called the Standard Model. It is clear to physicists – who have been working since the early twentieth century with two logically distinct theories, Relativity and Quantum Mechanics – each of which admirably serves scientific purposes, including the building and operation of complex machinery – that, however true their theories are, they are unlikely to be the last word.

Relativity and Quantum Mechanics are, for all present practical purposes, (reasonably) true theories. But that is because they have been made up by scientists to account for, and to operate on, things we want to understand and manipulate. Hacking gives an excellent example of how this works in his discussion of the Hall effect in electromagnetics, explained by the law that an electrical current passing through a conductor at right angles to a magnetic field produces an electric potential:

> If anywhere in nature there is such an arrangement, with no intervening causes, then the Hall effect occurs. But nowhere outside the laboratory is there such a pure arrangement. There are events in nature that are the resultant of the Hall effect and lots of other effects. But that mode of description – the interaction or resultant of a number of different laws –

---

18   Ibid., p. 265. Extended to the characterisation of modern science, this formulation still puts too much emphasis on what might be taken as an attitude or 'method' of thought; as Beller reminds us, '[w]hat became distinctive of modern science was not so much the scientific "method", or scientific "norms", but the strongly institutionalized communication system (scientific periodicals, scientific gatherings) to handle disagreement – a system that one cannot afford to ignore if one is to survive as a professional in a scientific community' (*Quantum Dialogue* [1999], p. 213).

19   Newton, *Principia Mathematica* (1966), p. 400; cited by Easlea, *Witch Hunting* (1980), p. 180.

is theory-oriented. It says how we analyze complex events. We should not have the picture of God putting in the Hall effect with his left hand and another law with his right hand, and then determining the result. In nature there is just complexity, which we are remarkably able to analyze. We do so by distinguishing, in the mind, numerous different laws. We also do so, by presenting, in the laboratory, pure, isolated phenomena.[20]

Not only the law but occurrences of the Hall effect itself are constructed – the latter literally so, using lab equipment, following theoretical considerations. Especially because we can make the effect happen, we say the law is true.[21] The fact thus stated is dependent for its existence on the previous history of physics, which included the production of the idea of the 'interaction' of 'forces', and the construction of the equipment that allowed Hall to experience the effect named after him – which was not the one he was originally looking to find.[22] As one historian of science has put it, because '[w]hatever culture is now structures whatever it might become, ... future knowledge production is intimately and irrevocably tied to the culture in which it is produced and which it is produced from'.[23] This is not to say that a culture can make up whatever 'effects' it wants, as Hall discovered when he went looking for one thing and found another, but that it makes the facts it can.[24]

This human-made nature of natural facts is especially apparent if – to return to an example mentioned earlier – we look at the world as described by the four fields whose operations are currently held to produce all physical phenomena: the gravitational, electromagnetic, weak, and strong fields. As in the case of the interaction of the Hall effect with other forces, we can 'compute each of

---

20   Hacking, *Representing and Intervening* (1983), p. 226.
21   This point is made in a seemingly more radical way by Bruno Latour and Steve Woolgar, in terms of their understanding of laboratory equipment as 'inscription devices' (machines that transform natural processes into sets of numbers or linguistic descriptions), without which 'none of the phenomena "about which" [scientists] talk could exist ... Without a bioassay, for example a [certain biological] substance could not be said to exist ... Similarly, a substance could not be said to exist without fractionating columns, since a fraction only exists by virtue of the process of discrimination. Likewise, the spectrum produced by a nuclear magnetic resonance (NMR) spectrometer would not exist but for the spectrometer' (*Laboratory Life: The Social Construction of Scientific Facts* [1979], p. 64).
22   See Hacking, *Representing and Intervening* (1983), pp. 224–5.
23   Pickering, 'Knowledge, Practice, and Mere Construction' (1990), p. 702.
24   Thus, to cite an example from an acclaimed and controversial study, 'given their extensive training in sophisticated mathematical techniques, the preponderance of mathematics in particle physicists' accounts of reality is no more hard to explain than the fondness of ethnic groups for their native language' (Pickering, *Constructing Quarks* [1984], p. 413).

the component forces acting on a particle according to the relevant set of field equations and then use the Law of Total Force to determine the trajectory of the particle'. Nevertheless, the theories describing each field 'cannot be joined into a single theory of the physical world' and 'it is not even demonstrable that the conjunction is consistent'.[25] The logical distinctness of Relativity Theory and Quantum Mechanics is a result of the history of physics (so far), though the former makes GPS systems work and the latter explains the operations of the transistor-based computer. In such a situation the contributions of the human subject and the natural object of knowledge are intimately interrelated.

In her fascinating anthropology of a biochemistry laboratory, Karin Knorr-Cetina began, much as I concluded my critique of philosophies of social science, 'with neither subject nor object, but with the concept of scientific practice'. The world of objects and their interactions constituting the phenomena studied by the sciences

> displays itself as an upshot of this scientific practice, meaningful and relevant only *within* the social constitution we have characterized, but at the same time not locked into subjective conditions. For it is precisely the selective constitution of scientific objects which is negotiated, imposed and deposed in this practice and which is itself at stake in the discourse crystallised in scientific operations.

For this reason she was led 'to blur the increasingly popular distinction between the natural or technological sciences on the one hand, and the social or cultural sciences on the other'. Both are the products of social-historically situated human beings engaged at once in the production of knowledge and the interpretation and symbolic shaping of experience.[26] The distinction between the two domains of science might then – as I suggested in the last chapter of the present book – be understood as a feature specific to the modern, capitalist culture in which the current idea of 'science' itself developed.

In the case of Marx's theory the laboratory construction of working models of selected elements of complex phenomena – the mainstay of contemporary natural sciences – is not possible: For instance, the workings of the law of value cannot be examined in isolation from all the factors that alter its effects in the economic system, such as the existence of unproductive capital and credit mechanisms. As Marx put it in the Preface to *Capital*, 'in the analysis

---

25   Joseph, 'The Many Sciences and the One World' (1980), p. 783.
26   K.D. Knorr-Cetina, *The Manufacture of Knowledge. An Essay on the Constructivist and Contextual Nature of Science* (Oxford: Pergamon Press, 1981), pp. 136–7.

of economic forms neither microscopes nor chemical reagents are of assist-
ance. The power of abstraction must replace both'.[27] Marx's theory therefore
begins with the construction of an idealised model of the system, tested by
comparison of its implications with phenomena observable in the course of
the evolution of capitalist society (such as recurrent crises and the tendency to
the displacement of labour by machinery in production processes).[28] In this his
modus operandi is, as Marx himself asserted, no different from that successfully
employed by the natural sciences.

The object investigated by Marx, capital, itself incorporates social constructions – notably, the quantification of human labour time and its representation
by sums of money – but, I argue in *Social Knowledge*, there is no inherent
reason why social constructions cannot be studied in the same way as natural phenomena – which as Knorr-Cetina suggests are themselves only known
in the form of social constructions. In the three decades since *Social Knowledge* was published, I have found no reason to doubt this conclusion. The track
record of economic science remains as dismal as it was when I wrote the book,
its most notable recent failure being the surprise of almost all experts at the
eruption of financial crisis in 2007, compounded by their inability to account
in a way generally agreed on for the ensuing global stagnation. On the other
hand, Marx's theory continues to provide convincing explanations for otherwise unexplained phenomena, such as the failure of Keynesian techniques to
abolish the business cycle or the inability of a relatively uncontested capitalist
economy to avoid crisis and stagnation. As is to be expected in a period of economic difficulty, interest in Marxist concepts has grown in recent years, along
with a generalised lack of confidence in the future of capitalism. However,
while the social forms traditionally studied by anthropologists have continued
to disappear with the accelerating destruction of non-capitalist modes of life,
the anthropological nature of Marx's investigation of contemporary society has
yet to be widely understood, and he is still commonly thought of, by admirers
as well as critics, as himself an economist.

On the other hand, my excitement at belatedly encountering the writing of
Pierre Bourdieu involved an appreciation of his monumental efforts to apply
the techniques of anthropological analysis to the study of his own culture –
indeed, to his personal milieu, the French university system, and the culture
of his own social class. Bourdieu's studies of art, literature, and intellectual life,
like Marx's critique of economics, demonstrate the possibility of the scientific

---

27   Marx, *Capital*, Vol. I (1976 [1867]), p. 90.
28   For a discussion of this modelling procedure, see my *Theory as Critique* (2018), Ch. 2.

study of one's own social reality; like Marx he uncovers contradictions inherent in modern social practices and explains them in terms of the dynamic power relations between social groups. By turning the attention of social science to its own practitioners he sought to maximise the sceptical reflexivity fundamental to the scientific attitude in general. The object of study, as he puts it, must be 'not the lived experience of the knowing subject, but the social conditions of possibility, and therefore the effects and limits, of this experience and, among other things, of the act of objectivation'.[29]

While Bourdieu's analysis of the artistic field and its peculiar place in the modern social order provided a foundation for my own studies of the ideology of art in contemporary society,[30] his overall project – the historical ethnology of modern society – reinforced my understanding of Marx's accomplishment in the critique of economic ideology. Thanks to Bourdieu's work, as well as to much of the history, sociology, and philosophy of science produced by the intellectual upheaval of the late 1960s, *Theory as Critique* is a better book than I could have written in 1986. But I could not have written it without first working through the problems treated in *Social Knowledge*, and I hope that others also will find reading this earlier work worthwhile.

November 2017

---

29   Bourdieu, *Science of Science and Reflexivity* (2004), p. 93.
30   Mattick, *Art in Its Time* (2003).

Preface

This book is one of several published in recent years dealing with issues in the philosophy of social science so basic as to call into question the very possibility of such a form of knowledge. It differs from most of these books in that it examines some of these issues with an eye to their appearance in the context of a particular theory, Marx's critique of political economy. This reflects both my interest in this theory for its own sake, and my feeling that it is at least in part the absence of discussion of the problems raised by actual social theory that gives much philosophy of social science its peculiarly sterile flavour.

On the one hand, arguments for the impossibility of scientific study of one or another realm of phenomena are in general dubious. The development of new methods may always open hitherto closed areas to scientific analysis. On the other hand, proofs of the mere possibility of a science (as exemplified by Ernst Nagel's defence of the social sciences in *The Structure of Science*) can be nearly as uninteresting. Marx's *Capital*, meanwhile, contains the statement of a theory constructed on a sufficiently large scale to be comparable to interestingly general theories in other domains. It meets the main requirements for a scientific theory: on the formal side, it proposes a set of principles for the explanation of observable realities and the prediction of definite trends; it is falsifiable, though it is in fact quite well confirmed. This example, however, has only rarely been discussed in the literature of the philosophy of social science. To begin with, Marx's work is not recognised by the mainstream of academic social science with respect to his own estimate of its importance: as a theory of capitalist accumulation and crisis. Even among those social scientists who style themselves radical or Marxist, the majority do not accept Marx's own theory as an accurate representation of social reality. From the point of view of the philosophy of social science this is doubly unfortunate: not only because Marx's work represents the outstanding example of a social theory, but also because it includes, as part of its basic framework of ideas, explicit attention to problems of interest to philosophers – while raising a new one: the curious inability of social theorists to accept the validity of Marx's analysis of capitalism despite its remarkable scientific strength.

My own interest in the philosophical questions with which this book is concerned developed while teaching college courses on *Capital*. The greatest barrier to students' understanding Marx's ideas, I discovered, was their difficulty in seeing these ideas as formulated from a point of view outside of the academic disciplines of economics, sociology, or political science. The basis task I faced was to explain what Marx meant when he described his theory as a critique of political economy rather than as a contribution to economics. Explaining this involved discussing Marx's conception of economics as not just a theory but a systematisation of ideas that help define as well as explain the forms of behaviour constituting capitalist social life.

In this way I was led to the classic anthropological problem of the relation between the cultural insider, the native for whom reality is defined by culturally developed forms of experience and thought, and the outsider, the scientific investigator who wishes both to grasp and to explain the native's way of life. The problem is posed to the extent that there is conflict between the anthropologist's and the native's understanding of the latter's world: what, if anything, can justify the scientist's claim to provide an explanation of native customs superior to the native's own, while doing justice to the role played by culture in the very construction of experience as well as in its comprehension?

I found it useful to introduce Marx's anthropological treatment of his own culture with a short discussion of E.E. Evans-Pritchard's analysis of Azande witchcraft beliefs and practices. Just as Evans-Pritchard attempted to explain Azande ideas and rituals in terms of their place in native social life, Marx wished to explain the continued faith in economics – despite its striking weakness as a science – displayed by the natives of his (and our) own culture in terms of the central role played in capitalist society by this system of ideas. This comparison proved helpful for my students; it led me to the questions, about the nature of scientific thinking and its relation to our everyday knowledge of social reality, and about the nature of that reality itself, discussed in this book.

While I believe that Marx's theory provides a starting point from which these questions can be answered, the present essay is neither an introduction to nor a full-scale discussion of Marx's conception of science and its realisation in his work.[1] The basic question it confronts is an abstract one: how is scientific knowledge of social life possible? That an answer to this question does not merely depend on general philosophical considerations but requires reference to the particular historical circumstances of the development of social theory will emerge from simultaneous consideration of a sister question: why does such knowledge seem impossible, despite some two hundred years of intellectual effort? My argument, in a few words, is that the difficulties of the social sciences have been due not to the inherent resistance of social life to scientific explanation, but to the culturally determined inability of would-be social scientists to subject their own categories for social experience – those of capitalist

---

1   [For a discussion, more than introductory but less than complete, see my *Theory as Critique* (2018), especially Chs. 2 and 11.]

society – to the cross-cultural comparison on which the possibility of scientific understanding of social life depends. It was by his ability to look at capitalism from the perspective of its eventual abolition that Marx succeeded both in explaining the limits of bourgeois social theory and in constructing a scientific alternative.

∴

Even so short a book as this one is a product of the intellectual work of many people besides the author. My thinking on the topics here explored reflects years of discussion with other members of the group Root & Branch, and especially with Elizabeth Jones Richardson. By inviting me to teach with him, Fred Moseley made possible the shared study of *Capital* in the course of which the comparison of economics and witchcraft first occurred to me. Hilary Putnam supervised the dissertation from which this book has developed, in the process kindly discouraging me from chasing a number of philosophical red herrings. Rochelle Feinstein and Ilse Mattick read the penultimate draft with care; their suggestions led to many improvements. Finally, anyone who knows his writings will recognise how much I owe to my father, Paul Mattick. From him I learned not only how to understand Marx but why it is important to do so: that the point of understanding the world really is to change it.

The book is dedicated to my teacher and friend Frans Brüggen.

CHAPTER 1

Introductory

In his preface to the second edition of *The Structure of Scientific Theories*, Frederick Suppe expresses confidence that the 'confusion and disarray within the philosophy of science' of the previous decade 'is becoming resolved' and has given way to a relatively coherent state of the field.[1] However true this may have been, the same certainly neither could nor can be said for the awkward subtopic of Suppe's field, the philosophy of social science. In the words of an author of a recent survey,

> The initial impression one has in reading through the literature in and about the social disciplines during the past decade or so is that of sheer chaos. Everything appears to be 'up for grabs'. There is little or no consensus – except by members of the same school or subschool – about what are the well-established results, the proper research procedures, the important problems, or even the most promising theoretical approaches to the study of society and politics.[2]

This statement only reflects a striking degree of agreement on the unsatisfactory state of social theory itself, not only among philosophical outsiders but among stock-taking anthropologists, sociologists, economists, and political scientists themselves. Although economics is often touted as the best developed of the social sciences, it is freely admitted by respected practitioners that its more theoretical areas have little connection with economic reality, while relatively empirical efforts, such as economic forecasting, also bear little impressive fruit.[3] In Tjalling Koopman's words, 'after more than a century of intensive

---

1 Suppe (ed.), *The Structure of Scientific Theories* (1977), p. v.
2 Bernstein, *The Restructuring of Social and Political Theory* (1978), p. xii.
3 Leonard Silk, writing in *The New York Times* of 27 July 1983, reported recent research on economic forecasting, by Victor Zarnowitz of the National Bureau of Economic Research:
     After examining the forecasts of 79 individual economists and firms made between the fourth quarter of 1968 and the first quarter of 1969, Mr Zarnowitz found that 'no single forecaster has been observed to earn a long record of superior overall accuracy, and indeed nothing in the study would encourage us to expect any individual to reach this elusive goal.
     'For most of the people, most of the time', he found, 'the predictive level is spotty, but with transitory spells of relatively high accuracy', a finding that might apply to the gambling tables at Las Vegas or Atlantic City ...

© KONINKLIJKE BRILL NV, LEIDEN, 2020 | DOI:10.1163/9789004414822_002

activity ... [one] is led to conclude that economics as a scientific discipline is still somewhat hanging in the air'. He is forced to plead the case for his discipline on the basis of its unknown *future*, in a statement so striking as to deserve quotation at length:

> If overestimation of the range of validity of economic propositions is the Scylla of 'informal' economic reasoning, a correct appraisal of the limited reach of existing economic theory may cause us to swerve into the Charybdis of disillusionment with economic theory as a road to useful knowledge. The temptation to identify the results of existing economic theory with economic theory as such [*sic!*] – and to disqualify both in one breath – is strongest for the experienced economic advisor in government or business, to whom the limitations of existing theory are most painfully apparent.[4]

Tom Bottomore, in his UNESCO-sponsored 'guide to the problems and literature' of *Sociology*, expressed the general opinion when he wrote that 'there is not, at the present time, any general body of sociological theory which has been validated or generally accepted'.[5] This remark would, I think, be widely accepted as applicable to anthropology, demography, political science, and history (for those who classify this field as, or in part as, a science). In general, as Bottomore says, 'one powerful argument against the scientific character of the social sciences has been that they have not in fact produced anything resembling a natural law'.[6]

The economist's strategy, if he is to gain fame and higher rewards for his ability to beat the standard odds, must be to capitalize on what Mr Zarnowitz calls the 'transitory spells of relatively high accuracy' by taking risks in the first place and then heavily advertising his occasional triumphs to keep them in public consciousness.

The risk-taking forecaster has the advantage of dealing with a public that seems to have a very good memory for forecasting success and a very poor one for failures. This nonsymmetrical public memory appears to be correlated with general credulousness and greed. As Phineas T. Barnum, the great American promoter and showman put it, 'There's a sucker born every minute.'

See also the works cited in Chapter 5, note 63. [This was written at a time when many companies still had forecasters on their staffs. By the first decade of the twenty-first century, 'thanks to the poor historical performance of economic forecasting, almost none of the Fortune 500 companies directly employ economists. Instead, they avoid relying on forecasts altogether ...' (Knoop, *Recessions and Depressions* [2004], p. 125).

4   Koopmans, 'The Construction of Economic Knowledge' (1968), pp. 537–8, 541.
5   Bottomore, *Sociology: A Guide to Problems and Literature* (1971), p. 29.
6   Ibid., p. 32.

Some take this state of affairs as an indication of the 'youth' of the social studies, relative to (say) physics, but posit a body of objective laws as the goal to which these studies are well on their way. Others see the lack of success in reaching this goal as representing something more serious, indeed as demonstrating the fundamentally misguided nature of the current idea of a social science itself. The question as to the very possibility of social science has roots far deeper than the insufficiencies of actual work in these areas: ultimately in the only slowly dissipating difficulty felt by European culture with treating human beings, and their social interaction, as part of nature. This difficulty has declined in physical and social anthropology since the day of Darwin's struggle against the idea of man's essential separateness; it is weakening further with the growing acceptance of biological approaches in psychology; but in the realm of social theory the accessibility of human phenomena to scientific thinking remains a topic for debate.

As May Brodbeck observes in the introduction to her reader in *The Philosophy of the Social Sciences*, the modern idea of such sciences can be traced to the thinkers of the Enlightenment. 'Seeing men as part of the natural order, they envisioned a science of man and society modeled on Newton's explanation of heaven and earth, by whose explanation the potentialities of man could be realized to form a more just and humane social order'.[7] It has remained a guiding assumption in this tradition of social theory that it is possible to discover general laws of social behaviour, explanatory of observed phenomena. And though it has been hoped that such laws would serve as guides to policy, believers in the possibility of such sciences have also emphasised the necessity and possibility of 'value-freedom' in research and analysis in these realms. That is, while the subject matter of the social studies is such as to inspire strong feelings and valuations – such, in short, as to express a point of view or way of life – this tradition has held that these feelings and values must and can be put aside in order to attain results as objectively valid as those of the physical and biological sciences.

A line of descent of equal age, however, stems from Vico's distinction between human and natural history: that we may have special knowledge of the former as it is we who make it. There are many versions current of the idea that consideration of the human realm demands special modes of inquiry, radically different from the methods appropriate to natural science. In general, this line of thought involves some form or another of the claim that

---

7 Brodbeck (ed.), *Readings in the Philosophy of the Social Sciences* (1968), p. 1.

> what is needed above all is a way of looking at social phenomena which
> takes into primary account the intentional structure of human conscious-
> ness, and which accordingly places major emphasis on the meaning social
> acts have for the actors who perform them and who live in a reality built
> out of their subjective interpretation.[8]

People have not only made their worlds: they make and remake them. This
radically 'subjective' aspect of human phenomena poses (or is held to pose)
all sorts of problems for scientific research. It is, for instance, supposedly not
the scientist's task to advocate normative positions. But since humans are
self-interpreting beings differences in interpretation of social experience – as
between a social group and a scientist investigating it – involve differences in
the ways in which the potential data for a social theory are constructed. On a
relatively simple technical level, this problem appears as that of the interfer-
ence of the researcher's values with those of the objects of his or her study. A
more fundamental objection, to the very possibility of social science, maintains
that social life can only be understood from 'within' – that is to say, in terms of
the reasons and values that define what the actions constituting a way of life
are, to begin with. If the student of society is 'outside' the culture investigated,
he or she will not be able to understand it; on the other hand, if he or she is
'inside' it, it is difficult to see what meaning can be given to the 'objectivity'
associated with the idea of science.

One reason the methodology of the social sciences is so well populated is the
fact that these two lines of descent have always been intimately intertwined –
if not incestuously, at any rate by every variety of cross-cousin marriage. No
one, for example, has insisted more forcefully than Max Weber on the necessity for value-freedom and scientific objectivity in the social sciences. Yet it was
he who described *Verstehen*, 'interpretive understanding' of subjective points
of view, as a fundamental method of these fields. Similarly, Ernest Nagel has
argued that one can both admit the subjective nature of the stuff of social analysis and hold that 'the logical canons employed by responsible social scientists
... do not appear to differ essentially from the canons employed ... in other areas
of inquiry'.[9]

This attempted synthesis of objectivity and subjectivity, I think, charac-
terises the majority opinion within the social sciences today: such sciences
are supposed to produce objective knowledge of the causes, nature, and con-

---

8 Natanson, 'A Study in Philosophy and the Social Sciences' (1963), p. 273.
9 Nagel, *The Structure of Science* (1961), p. 484.

sequences of the complex structures of subjective meaning that constitute social phenomena. Weber's definition is classical:

> Sociology ... is a science which attempts the interpretive understanding of social action in order to arrive at causal explanation of its course and effects. In 'action' is included all human behavior when and in so far as the acting individual attaches a subjective meaning to it.

The idea of 'subjectivity' here puts emphasis on the individual as the basic unit of analysis, whose 'action is social in so far as, by virtue of the subjective meaning attached to it by the acting individual ... it takes account of the behavior of others and is thereby oriented in its course'.[10] An example of such 'interpretive understanding' would be the explanation of economic behaviour in terms of the 'profit motive' of entrepreneurs.

In contrast, Marx's social methodology appears to champion 'scientism' in its purest form. What greater contrast to Weber's attempted synthesis of subjective and objective could be found than Marx's declaration in the Preface to *Capital*, Volume I, that

> Individuals are here dealt with only in so far as they are personifications of economic categories, the bearers of particular class-relations and interests. My standpoint, from which the development of the economic formation of society is viewed as a process of natural history, can less than any other make the individual responsible for relations whose creature he remains, socially speaking, however much he may subjectively raise himself above them.[11]

Marx's attitude is explicitly connected with his intention to construct a science of society on the model of physics and chemistry. Beside the passage just quoted we may set the statement of his 'standpoint' which is one of the best-known passages from Marx's works:

> In the social production of their existence, men inevitably enter into definite relations, which are independent of their will, namely relations of production appropriate to a given stage in the development of their material forces of production. The totality of these relations of produc-

---

10   Weber, *The Theory of Social and Economic Organization* (1947), p. 88.
11   Marx, *Capital*, Vol. I (1976 [1867]), p. 92.

tion constitutes the economic structure of society, the real foundation, on which arises a legal and political superstructure and to which correspond definite forms of social consciousness. The mode of production of material life conditions the general process of social, political, and intellectual life. It is not the consciousness of men that determines their existence, but their social existence determines their consciousness ... [C]hanges in the economic foundation lead sooner or later to the transformation of the whole immense superstructure. In studying such transformations it is always necessary to distinguish between the material transformation of the economic conditions of production, which can be determined with the precision of natural science, and the legal, political, religious, artistic or philosophical – in short, ideological – forms in which men become conscious of this conflict and fight it out. Just as one does not judge an individual by what he thinks about himself, so one cannot judge such a period of transformation by its consciousness, but on the contrary, this consciousness must be explained from the contradictions of material life, from the conflict existing between the social forces of production and the relations of production.[12]

This prescription seems in conformity to the work, the critique of political economy, to whose first publication it served as prefatory matter. Does not Marx display a model of capitalist society in which economic forces compel a developmental history in which people are caught up and which thrusts ideological views and modes of social action upon them? Even an historian like E.P. Thompson, who claims a Marxian heritage, protests against what he considers the exclusion of subjectivity from *Capital*, judging that work to be 'a study of the logic of capital, not of capitalism', for 'the social and political dimensions of the history, the wrath, and the understanding of the class struggle arise from a region independent of the closed system of economic logic'.[13]

At the same time, we should not forget, as Thompson does, that *Capital* is presented not as an economic theory, but as a critique of economics, itself described by Marx as 'superstructural'. If we look at Marx's actual analysis of capitalist society, a picture seemingly in conflict with his methodological prescription emerges. *Capital* begins with the analysis of commodities and so of

---

12   Marx, *A Contribution to the Critique of Political Economy* [1859], in Marx and Engels, *Collected Works* (henceforth MECW), Vol. 29 (1987), p. 263.

13   Thompson, *The Poverty of Theory* (1978), p. 65.

commodity exchange. The latter is not, as may at first appear, to be identified with the transfer of objects from one person's hand to another's. The owners of commodities, rather,

> must *place themselves* in relation to one another *as* persons whose will resides in those objects, and must behave in such a way that each does not appropriate the commodity of the other, and alienate his own, except through an act to which both parties consent. The [exchangers] must therefore *recognize* each other *as* owners of private property ... The content of this juridical relation ... is itself determined by the economic relation. Here the persons exist for one another merely as representatives, and hence owners, of commodities ... [I]t is as the bearers of these economic relations that they come into contact with each other.[14]

The last sentence establishes phraseological continuity with the passage quoted from the Preface to *Zur Kritik*. And yet the overall meaning seems to be quite different. Marx's idea is that the modern concept of property, as defining a certain social point of view and rule for behaviour, is necessary for capitalist market exchange to take place (since, for example, collectively owned products cannot be exchanged within the collective). The economic relation is fundamental in the sense that the exchange-relation is involved in the definition of 'property' (since a good's being one's property here implies one's having the right to sell it). But at the same time the exchange-relation itself is described with reference to the kind of phenomena commonly called 'subjective' in the social science literature. It is only because they *treat* each other *as* private proprietors that the individuals concerned can be 'bearers' of this economic relation. In fact, this formulation seems not incompatible with Weber's definition of 'social action', cited above. The economic relation depends upon the existence of the juridical relation, and the latter clearly involves an element, at least, of 'consciousness': at any rate we are here within the 'superstructure' which the economic 'foundation' is supposed to determine!

With this we touch, of course, on one of the classic problems of Marxism. By bringing this problem of the relation of basis to superstructure into contact with the bourgeois social theorists' discussion of the relation between subjective and objective aspects in the construction of social reality, I wish to illuminate the issues raised in both traditions. I will begin with an examination of the concepts of 'subjective' and 'objective' as they have been developed in modern

---

14   Marx, *Capital*, Vol. I (1976 [1867]), pp. 178–9. My emphases.

social theory. Like many others, I will use Peter Winch's critique of Weber as a starting point. Despite its serious weaknesses, Winch's work remains valuable for its isolation of the central issues in modern social theory. I will try to show that the category of 'subjectivity' has maintained its place in this context largely thanks to a confusion of the 'intentional' character of social phenomena with the idea that such phenomena are unobservable. Having shown that the latter idea rests on a misunderstanding of the nature of scientific observation, I will turn in Chapter 3 to an examination of Winch's argument that social knowledge cannot be objective because the criteria for the truth of sociological propositions are always defined as relative to the culture in question. Here I will turn to another standby of the philosophical literature on the methodology of social science, Evans-Pritchard's study of Azande witchcraft, to argue that this is an untenable position. Chapter 4 will sketch an alternative conception of social understanding, and in Chapter 5 I will return to Marx's attempt to formulate methodological presuppositions for a scientific study of social life in such a way as to do justice to its intentional aspect. Finally, I will examine some important consequences of these presuppositions for our understanding of social science and its relation to other forms of social knowledge.

CHAPTER 2

'Subjective' and 'Objective'

According to F.A. Hayek, 'there are no better terms available to describe [the] difference between the approach of the natural and the social sciences than to call the former "objective" and the latter "subjective." Yet these terms are ambiguous and might prove misleading without further explanation'.[1] On the one hand, he says, the facts of social phenomena are as objective as natural facts, in the sense that they are given independently of the observer, and are thus public, given equally to every observer. On the other hand, the social sciences are oriented towards subjectivity in that they 'deal in the first instance with the phenomena of individual minds, or mental phenomena, and not directly with physical phenomena. They deal with phenomena which can be understood only because the object of our study has a mind of a structure similar to our own'. In addition, 'the term "subjective" stresses ... that the knowledge and beliefs of different people, while possessing that common structure which makes communication possible, will yet be different and often conflicting in many respects'. Thus, finally, the social sciences may be described as 'concerned with man's conscious or reflected action, actions where a person can be said to choose between various courses open to him ...'.[2]

I choose Hayek as a starting point not because his analysis is particularly clear or persuasive, but because it includes – without clearly distinguishing them – the chief uses of the pair 'subjective/objective' to be found throughout the literature of and about the social sciences. Three uses of this pair of terms can be distinguished in the passages just quoted. (1) First of all, social facts are held to be objective relatively to the observer's subjectivity, in the sense that they exist as objects for the observer's consciousness, rather than as its products. (2) However, says Hayek, if we look at these objects of social study we see that they themselves have a subjective character, in that their existence depends on *some* persons' minds – the minds of the persons studied. For this reason the facts relevant to social theory are, 'in the first instance', those which exist for the members of the social group under investigation. One result is that social phenomena exist distinctively in a world defined by choice and valuation, rather than by natural causality. (3) Finally, Hayek stresses the individual

---

1 Hayek, *The Counter-Revolution of Science* (1955), p. 28.
2 Ibid., pp. 28, 29, 26.

© KONINKLIJKE BRILL NV, LEIDEN, 2020 | DOI:10.1163/9789004414822_003

as the basic unit of the social world. Society's stuff is subjective because it is composed of individuals and because no two individuals are alike.

This third use of 'subjective' will not be dealt with here. It has played an important role in various 'individualist' formulations of social theory, and there is a large literature devoted to the topic.[3] In general, for obvious reasons the individuals dealt with by social theories have not been real, particular people, but ideal representatives thereof, defined either in terms of a putative human nature or in terms of the institutions, norms, etc. of a given social situation. This is clearly to be seen in the case of the classical concept here, Max Weber's 'ideal type' or 'the theoretically conceived *pure type* of subjective meaning attributed to the hypothetical actor or actors in a given sort of action'.[4] (An example is the ideal type of the entrepreneur in capitalist society, supposedly motivated purely by economic gain.) Models of social action using such ideal types are intended to yield approximations to, and so explanations of, real social behaviour, in which the actual individuals are much more complexly motivated. As is already indicated by the phrase 'ideal type', Weberian individualism (like methodological individualism generally) amounts to little more than the recognition that society is made up of individuals; the actual burden of explanation is carried by the type concepts whose definition requires reference to social relations between persons.[5] One could even say that the 'ideal individual' of social theory has been a conceptual representation of what anthropologists call a 'culture' – a system of learned patterns of life shared by members of a society – as embodied by the individuals who have learned it. We will return to the concept of culture and its relation to the 'subjective' below.

Since Hayek's first use of 'subjective/objective' merely indicates a minimum condition for use of the word 'knowledge' – that there be an object to be known in some sense independent of the knower – the rest of this chapter will focus on Hayek's second use of these terms. To begin with, he indicates, if we wish to explain human behaviour with respect to any natural phenomena, the latter must 'not be defined in terms of what we might find out about them by the objective methods of science, but in terms of what the person acting thinks about them'. Hayek elaborates this in two ways. First of all, he claims, 'any knowledge which we might happen to possess about the true nature of the material thing, but which the people whose action we want to explain do not possess,

---

3 For a representative selection of articles on methodological individualism, see Brodbeck (ed.), *Readings* (1968), Chs. 13–18; for a critique see Garfinkel, *Forms of Explanation* (1981).
4 Weber, *Theory* (1947), p. 89.
5 See the editors' introduction to Gerth and Wright Mills (eds), *From Max Weber: Essays in Sociology* (1946), pp. 57–8.

is as little relevant to the explanation of their actions as our private disbelief in the efficacy of a magic charm will help us to understand the behavior of the savage who believes in it'.[6] We will spend most of Chapters 3 and 4 dealing with a version of this claim, for, unlikely as it may seem, it has been advanced as basic to the nature of social knowledge. Interestingly, however, this point has been developed by critics of the 'mainstream' social science of which Hayek is a representative, such as Peter Winch.

We shall here deal only with the more conservative formulation of the contrast between 'subjective' and 'objective', as represented by Weber's assumption that in addition to the subjective side of a social practice there are associated 'processes and phenomena which are devoid of subjective meaning', but of which account must be taken in social explanation. Admitting that the realities of physical nature play their role in social life, Weber is concerned to stress that this role is shaped by the imposition on nature of human-determined meanings.[7] Hayek presents a similar view, although he (typically) does not note its difference from its more radically subjectivist cousin. In his words,

> Neither a 'commodity' or an 'economic good', nor 'food' or 'money' can be defined in physical terms but only in terms of views people hold about things ... Nor could we distinguish in physical terms whether two men barter or exchange or whether they are playing some game or performing some religious ritual. Unless we can understand what the acting people mean by their actions, any attempt to explain them, i.e., to subsume them under rules which connect similar situations with similar actions, are bound to fail.[8]

This passage makes two different points. First, it observes that socially significant phenomena are culturally and not (just) physically defined: what one group accepts as food, for instance, may be rejected as inedible filth by another. Second, Hayek extends the point into a rejection of what is generally called 'behaviourism' as a method for social science, arguing that to understand human action requires not just what he calls 'physical' information – e.g. observations of body movements – but also information about what the human subjects mean by their actions.

It is important to see that these are two distinct points, however closely related they may be in the minds of the majority of writers on social theory. The

---

6 Hayek, *Counter-Revolution* (1955), p. 30.

7 Weber, *Theory* (1947), p. 93.

8 Hayek, *Counter-Revolution* (1955), p. 31.

first only specifies something to which we will give the vague name 'culture' as the subject matter of the social sciences. The second involves a certain view of the relation between 'the mental' and 'the physical', and so an interpretation of those concepts themselves. The latter concept is identified, as the label indicates, with the domain of phenomena open to sense perception, while these in turn are taken to be none other than the objects describable by categories of physical theory: categories of motion through space-time. Events in this domain are to be explained by reference to *causal* laws ('rules which connect similar situations with similar actions', in the case of social events). 'The mental', in contrast, is the world of phenomena unobservable with the techniques of natural science but understood by way of the mind-to-mind communication of meanings, and explained typically by reference to the *reasons* given for actions.

Here again Max Weber makes carefully explicit what Hayek suggests when he states the difference between the mental and the physical in terms of the idea of 'rationally purposive action'. The distinguishing feature of such action in Weber's conception (as referred to in the definition of sociology quoted above) is that it is behaviour regulated by a conscious agent in accordance with some purpose of the agent's – in Weber's words, 'a subjectively understandable orientation of behavior'.[9] To understand it requires interpretation of the action in terms of the grounds that the agent had, or can be seen as having had, for his action. This, what Weber called explanatory 'adequacy on the level of meaning', he contrasted with 'causal adequacy' of interpretation, which is achieved by the subsumption of a given action under an empirical generalisation.[10]

The aim of social science, for Weber the 'correct causal interpretation' of action, is achieved when a (typical) action-process is 'both adequately grasped on the level of meaning and at the same time the interpretation is to some degree causally adequate'.[11] The action, that is, must both be interpreted as rationally motivated and be subsumed under a statistical law linking occurrences of a certain situation – say, increasing demand for a good – with the action in question – say, raising its price. Both parts of the explanation are necessary, in Weber's view, since it is only understanding the means-end structure of an action that makes possible explanation of it as an action, as opposed to instinctive or reflexive behaviour, while only empirical generalisations yield causal laws. Social explanation thus requires the combination of two methods in the human sciences. To the empirical mode of study shared with the natural

---

9   Weber, *Theory* (1947), p. 101.
10   See ibid., p. 99.
11   Ibid.

sciences must be added the operation of *Verstehen*, interpretive understanding, in which the investigator attempts to reconstruct the motivational structure of the actions he is studying. The first studies actions as 'objective', the second as 'subjective': in this way the claims of science and those of the special nature of the social subject matter are both satisfied.

But can 'subjective' and 'objective', once distinguished in this way, be so smoothly combined in one explanatory framework? Weber's claim is that *Verstehen* is as much a scientific procedure as the making of empirical generalisations. Its essence is not some form of intuitive 'putting oneself in another's place' – though this may have heuristic value as a means to *Verstehen* – but modelling the means-end calculation that makes a given action rational. Ernest Nagel has pointed out an obvious problem with Weber's idea:

> In discussing the adequacy of the method of *Verstehen* it is essential to distinguish between that method conceived as a way of *generating* suggestive hypotheses for explaining social action, and that method conceived as a way of *validating* proposed explanations ... [I]t is generally recognized that the method of *Verstehen* does not, by itself, supply any *criteria* for the validity of conjectures and hypotheses concerning the springs of human action.[12]

Weber himself emphasised that 'verification of subjective interpretation by comparison with the concrete course of events is, as in the case of all hypothesis, indispensable'.[13] Verification is here identified with what are assumed to be the basic procedures for testing hypotheses in physical science – checking against the sense-observation of events in space-time. Unfortunately, he added, such verification is really only possible in 'the few very special cases susceptible of psychological experimentation' – in cases, that is, when the model of the motivation of an individual's action can be checked under laboratory conditions. In most cases, 'there remains only the possibility of comparing the largest possible number of historical or contemporary processes which, while otherwise similar, differ in the one decisive point of their relation to the particular motive or factor the role of which is being investigated'.[14] (An example is provided by Weber's most famous study, in which the difference in outcome of two pre-capitalist situations is explained in part by the presence in one of the 'Protestant ethic'.)

---

12   Nagel, 'On the Method of Verstehen as the Sole Method of Philosophy' (1963), p. 264.
13   Weber, *Theory* (1947), p. 97.
14   Ibid.

Aside from the many difficulties inherent in such a method of historical analysis, one can question the very compatibility of Weber's idea of the empirical verification of subjective interpretation with his conception of the subjective/objective distinction. A large number of motivational models are compatible with a given course of behaviour, to begin with. So long as subjective interpretation and causal generalisation are treated as two separate, if parallel, modes of explanation, it is hard to see how one can serve as a check on the other.

One way out of this difficulty opens if we reject the distinction Weber makes between causal and rational explanation. If, that is, we allow that reasons may be causes of observable behaviour, the two forms of explanation could be integrated, and motivational models could be evaluated by their success in predicting and retrodicting action. That reasons can be construed as causes has, I think, been convincingly argued by a number of writers.[15] I will not go further into this question here, because this way out of his dilemma is not open to Weber. Close though it is to his own position, he is blocked from reaching it by the conceptual structure of his subjective/objective dualism.

This emerges clearly if we look at Weber's proposed method of comparative sociology, which involves the comparison of institutional (and motivational) complexes of different social systems. Such comparison requires the investigator to judge that institutions from different systems are the same or different, or that typical actions taken by members of different social groups are similar or different. But by what criteria are these judgements of similarity and difference to be made? Since the social character of action for Weber consists in the subjective meanings attached to it by members of a society, identification of social practices seems to require knowledge of the correlation between motivational structure and physical behaviour *prior* to the observation of the latter. But then such observations cannot serve to verify interpretations of the former. As Peter Winch has put this objection, in *The Idea of a Social Science*, 'someone who interprets a tribe's magical rites as a form of misplaced scientific activity will not be corrected by statistics about what the members of that tribe are likely to do on various kinds of occasion ...'.[16]

Winch is an interesting critic of Weber, because in essence he is attacking Weber on the basis of the latter's own concept of the subjective, although it is rephrased in terms drawn from linguistic philosophy. Social relations, in Winch's view, are structured by shared concepts, which are embodied in and

---

15   See, in particular, Davidson, 'Actions, Reasons, and Causes' (1963), pp. 685–700.
16   Winch, *The Idea of a Social Science* (1963), p. 113.

so constitute a language. Language is described, following Wittgenstein, as both the expression of and a condition for a 'way of life'. Language is a social phenomenon by nature, as meaning depends on language-users' adherence to rules, which of necessity must be shared and not idiosyncratic. Language is thus not only a medium of social life but is paradigmatic of social activity, which is generally defined by the rules that regulate forms of conduct. For this reason the nature of knowledge about society 'must be very different from the nature of knowledge of physical realities'. The latter is based on the recognition of causal regularities, the former on the understanding of meaning relations. Knowledge about society is like knowledge of a language in that one cannot know a language without knowing the linguistic rules that define membership in the social group that speaks it. If we leave meaning relations out and just describe what we see when we observe social life, we will leave out what represents sociality *par excellence*.[17]

Despite the difference in terminology, all this is (as Winch says himself) in fundamental agreement with Weber's emphasis on subjective meaning as definitive of social action, as well as with his analysis of subjective meaning in terms of norm-governed or purposive behaviour. Thus Winch's 'rules' correspond to Weber's 'typical motivation patterns' and to the 'norms' of post-Weberian sociology.[18] On the other hand, on this basis Winch condemns Weber for his attempt to show 'that the kind of "law" which the sociologist may formulate to account for the behavior of human beings is *logically* no different from a "law" in natural science' by showing that a motivationally comprehensible scene can also be given a purely 'objective' description. In this attempt, as Winch describes it,

> Instead of speaking of the workers in [a] factory being paid and spending money, he speaks of their being handed pieces of metal, handing these pieces of metal to other people and receiving other objects for them ... In short, he adopts the external point of view and forgets to take account of the 'subjectively intended sense' of the behavior he is talking about.

In fact, for Winch such a description is not a description of social behaviour at all; Weber 'does not realize that the whole notion of an "event" carries a different sense here, implying as it does a context of humanly followed rules which cannot be combined with a context of causal laws in this way without creating

---

17   Ibid., pp. 123, 88.
18   See ibid., pp. 49–50.

logical difficulties'.[19] For example, one can break a rule, or make a mistake in following one, but one cannot break a causal relation. Interpretations cannot be added to causal descriptions, for one cannot even describe a social situation without knowing the interpretation; and for the same reason observation of causal sequences provides no check on subjective interpretation.

Winch makes essentially the same point in his discussion of Vilfredo Pareto's attempt 'to treat propositions and theories as "experimental facts" on a par with any other kind of such fact' or, more generally, with Emile Durkheim 'to consider social facts as things'. Discussing the former, Winch comments that

> In a sense Pareto has not carried his empiricism far enough. For what the sociological observer has presented *to his senses* is not at all people holding certain theories ... but people making certain movements and sounds. Indeed, even describing them as 'people' really goes too far, which may explain the popularity of the sociological jargon word 'organism': but organisms, as opposed to people, do not believe propositions or embrace theories.[20]

One can say that Winch himself is not carrying things far enough here. After all, the senses are not presented with organisms either but – if they are presented with anything – with sense data. If these are perceived as attributes of physical objects, classifiable as organisms, why shouldn't the latter be classifiable as people, and indeed as people holding and discussing views?

There are really two different questions here, which Winch like many writers on these topics is collapsing into one: (1) how do we go about describing and explaining a point of view, whether an individual's or one shared by members of a culture? and (2) how can we tell what's in people's minds, when we can't observe such mental contents with our senses? The conflation of these two questions into one rests on a conception of *observation* – and this on a whole system of epistemological categories – that, despite their longevity, are severely flawed.

To begin with, observing is not done 'with the senses' in Winch's use of the words in any field of study, or indeed in daily life (as Piaget's studies of the child's construction of the categories of experience have shown). It is by now widely recognised by philosophers of science that such a view of observation is not applicable even where social scientists and their philosophical critics

---

19   Ibid., p. 117.
20   Ibid., pp. 109–10.

often imagine it to be pre-eminently the case: in physics, where the categories of observation are clearly not given by nature but have been historically constructed in the form of physical theory. This is obvious in the case of the observation of quarks and leptons, but, as E.H. Hutten points out in his book *The Ideas of Physics*, 'even the "direct" observation of a familiar object, say of a star, presupposes not only a long process of learning acquired by the individual but also the knowledge accumulated by mankind over thousands of years'.[21]

Given this, it would be perverse to limit sociological observation to phenomena describable with the constructs of ethology, rather than those developed in the study of society itself. Since we know that for the members of any social group studied 'behaviour' in the physicalist sense does not exist, but only behaviour within the conventions of a given culture, there is no need to pretend that the former can be an object of study for the scientific observer either. More generally, there is no reason to give the concept of observation in social investigation a content – the 'pure sense-observation of physical occurrences' – it has in no other field of study. What social analysts observe are precisely the things to which Winch draws attention, under the rubric of the 'subjective': business behaviour, thinking, kinship structures, value systems, etc. The important question is not *how does one know about things one cannot observe?* but rather *how does one observe things of this kind?*

The situation here may be usefully compared with the problem of 'radical translation' posed by W.V.O. Quine. This problem indeed is raised in the form of a story about an imaginary ethnological study. An anthropologist, encountering a people whose language he does not speak, attempts to learn their language and so their cultural concept-system. [The problem is that, in the absence of a culturally-neutral observation language, the language-learner must always be *interpreting* the utterances of the native speaker; with no direct access to the latter's mind, the former can only attempt to match speech and other behaviours to other perceivable events. But in this situation there are always multiple ways in which language can be matched up to the world. In Quine's famous example, is the native who utters 'Gavagai' when a rabbit hops by saying 'Lo, a rabbit' or 'Lo, an undetached rabbit-part', or even 'Let's go hunting!'? Over time, some of these hypothetical translations become more likely than others, but the translation manual as a whole will always remain underdetermined by the behavioural data.[22]]

---

21   Hutten, *The Ideas of Physics* (1967), p. 3. For a detailed account of observation in physics, see Shapere, 'The Concept of Observation in Science and Philosophy' (1982).
22   See Quine, *Word and Object* (1960), Ch. 2.

In a very interesting article, John Wallace has observed that this problem can be seen as motivated by two issues important for some time in philosophical thinking. The first derives from the view of our conceptual scheme as divided into two parts: the system of *physical* concepts, which is 'basic and covers everything'; and 'our system of semantic, psychological, moral, aesthetic, and social concepts', the *mental* system. The questions arise: how is the boundary between the two systems to be drawn? And: how are they related to each other? The second of these questions corresponds to the Weber-Winch problem, how 'observable behaviour' is to be correlated with motivational or rule-following subjectivity.

The second issue finding expression in the problem of radical translation is that of 'the danger of projecting onto others our own patterns of thought and action, our own wishes, interests, and plans, our own standards, criteria, ideals, categories, concepts, and forms of life'. This corresponds to the question: how is social activity to be 'objectively' understood? Just as both Weber and Winch conflate this question with that of the correlation of behaviour with meaning, the radical translation problem, as Wallace sees it, connects these issues by the way in which it visualises the anthropologist's position in the face of his data. Part of his 'already acquired conceptual scheme is immediately applicable', in the Quinean view, to the alien ground on which he finds himself

> without fear of falsification or distortion: the physical part. Application of the other part, the mental system, has to be worked out. In working out the application of mental concepts what we have to go on are descriptions in physical terms of the environment and of the behavior of the inhabitants ... We have marked off ... an area of description of human beings which (i) is immune to the danger of projection and (ii) provides an evidential basis for application of descriptions from our remaining, projection-prone stock.[23]

This is pretty easily transferable to Weber's account of the use of statistical generalisations as a control on subjective interpretations of social action;[24] Winch's position may be stated in its terms as the denial that the two systems, the mental and the physical, are applicable to the same phenomena.

---

23   Wallace, 'Translation Theories and the Decipherment of Linear B' (1979), pp. 112–13.
24   Weber is not a psychological behaviourist. Thus he would accept introspective reports as evidence for the evaluation of motivational models. In his view, however, except for the rare laboratory experiment direct psychological access is not possible and we must construct and evaluate our models from the 'outside'.

Wallace compares this general model of radical translation with an actual case: the decipherment of the Cretan script, Linear B. According to his account, in the practice of interpreting and learning about other people, physical and mental concepts are mixed up in single reports in such a way that the two systems cannot be treated as independent. In the interpretation of certain symbols in the Minoan script, the scholars working on the case took as 'immediately accessible evidence what for the philosophers has to be theory [supported by physicalist evidence], a semantical piece of the mental framework'.[25] Wallace concludes that the correct answer to the question, how the application of mental concepts is constrained by the application of physical concepts, is: 'The question presupposes a process – application of mental concepts on the basis of physical concepts – which does not exist'.[26]

To return from this digression to our topic, Wallace's point applies both to Weberian social science and to Winch's argument against it, as they share the same dualistic starting point. (Indeed, Winch criticises Weber from a viewpoint defined by a more consistent application of the latter's concept of the 'subjective'.) The staying power of the word 'subjective', despite the inappropriateness of its application to what all agree are inter-subjective, social concepts and norms, derives I think from the implicit reference to the supposed 'mental' portion of the conceptual scheme we have inherited from Descartes, as opposed to the 'objective' or 'physical' portion. Even more than the emphasis on the individual as the basic unit of social life, the continued use of 'subjective' and 'objective' indicates the maintenance of a radical splitting of the phenomena of human life from the rest of nature. This is visible, for instance, in the very concept of 'behaviour', which is taken to pick out the physical side of action as distinguished from and thus as observable separately from the mental activity that gives it meaning.

This is, however, not the only way human action can be approached. For example, from the point of view of Piaget's theory of cognitive development, the psychological behaviourist's 'stimulus' does not really exist except as developmentally constructed by the organism involved: so that description of a sequence of actions in terms of 'stimulus' and 'response' describes 'behaviour'

---

25   Wallace, 'Translation Theories' (1979), p. 123.
26   Ibid., p. 137. [In *Knowledge, Belief, and Witchcraft* (1997 [1986]), Barry Hallen and J. Olubi Sodipo work through a specific problem in 'radical translation' in some detail, to show that with respect to understanding an alien thought-system by way of translating its basic expressions into the researcher's native language, such 'translation is extremely difficult but it is not *so* radically indeterminate ... [T]here are guidelines to follow. There are measures to take which may provide a clearer indication of alien theoretical meaning in the language of translation' (p. 84).]

only via implicit reference to what the behaviourist considers 'mental' phenomena. Piaget's conceptual framework, by making this explicit – by consid-
ering intelligence as the self-regulation of the 'behaving' organism – deprives
'mind' and 'body' alike of their status as separate analytical objects.[27] Similarly,
some anthropologists have defined the concept of 'culture' in such a way as
to allow us to pose the problem of establishing a science of social life in other
terms than those of demonstrating a concordance between 'adequacy of meaning' and 'causal adequacy'. For example, Clyde Kluckhohn offers the following
definitions of 'culture':

> that part [of human life] which is learned by people as the result of
> belonging to some particular group, and is that part of learned behavior
> which is shared with others ... By 'culture' we mean those historically cre-
> ated definitions of the situation which individuals acquire by virtue of
> participation in or contact with groups that tend to share ways of life that
> are in particular respects and in their total configuration distinctive.[28]

We may then by extension speak of the society or group defined by its possession of such shared learning as 'a culture'.

Whatever the limitations of this conception of culture, such an approach
to the phenomena of social life does not suggest the problem of the relation
between objective and subjective that Weber's definition of sociology brings
with it. Kluckhohn, defining 'explicit culture' as 'all those features of group
designs for living that might be described to an outsider by participants in a
culture' or that might be observed by the outsider, is careful to say that

> To avoid confusion, it should be noted that the basic data from which
> the anthropologists abstract explicit culture encompass manifestations
> of 'feeling' and 'thought' and are in no sense restricted to objects and acts
> in the narrow behavioristic sense. In other words, 'explicit' does *not* draw
> the line which 'objective' is supposed to draw from the 'subjective'.[29]

Aside from the matter of behaviourism (which, as noted, does not characterise
Weber's position), what is important here is that elements of cultures describable as 'feeling' and 'thought' are to be thought of neither as privately indi-

---

27   See, for example, Piaget and Inhelder, *The Psychology of the Child* (1969), pp. 4–6.
28   Kluckhohn, *Culture and Behavior* (1962), pp. 25, 52. Note that 'learned behaviour' does not
     have here the meaning that it has in behaviourist learning theory.
29   Ibid., p. 63.

vidual nor as inward, but as public and in principle observable. An example that appears to support the Weber-Winch view of the matter will clarify the point. In a critique of Winch, Alasdair MacIntyre wrote:

> Suppose that a team of Martian social scientists is observing human behavior. What they are watching we should describe as chess-playing, but unhappily they lack the concept of a game ... They therefore do not discern the rule-governed character of the players' behavior, although they arrive at many statistical generalizations about the movement of small pieces of wood by human beings. What is it that they do not understand when they fail to understand these movements as a game of chess? They fail of course first of all to grasp the players' actions as distinct from their physical movements. It is not that they wrongly *explain* what is done; rather they fail to *identify* the actions which are to be explained.[30]

MacIntyre seems to be saying the same as Winch in the latter's criticism of Weber. But there is an important difference. As MacIntyre states it, the failure of the Martians stems not from their limiting themselves to the 'external point of view', from which they attend only to what is 'physically observable', but from their lack of categories that would make the *observation* of chess-playing behaviour possible. Because their culture lacks games, they cannot identify what they are seeing, and will be forced to describe it in terms available from their own culture. They are thus in the toils of the problem of projection, but this derives from the presence of cultural blinkers on theory and not from the 'subjective' nature of social phenomena.[31]

To summarise the argument so far, we have seen that free and easy use of the opposition 'subjective/objective' has allowed confusion of recognition of the *intentional character* of social phenomena with the idea that such phenomena are *unobservable*. This confusion depends (at least in part) on a reductionist analysis of observation as the passive registration and subsequent organisation of sensory givens. As this analysis is not tenable for any science, least of all the physical ones, there is no call to apply it to the social studies.

---

30   MacIntyre, 'A Mistake About Causality in Social Science' (1962), pp. 61–2. The non sequitur
      in MacIntyre's argument – lack of the concept of a game need not spell inability to recog-
      nise rule-following – does not affect his essential point.

31   The problem of formulating categories for the observation of 'subjective' phenomena can
      be seen in perhaps its purest form in archaeology, where the human constructors of mean-
      ing have long since vanished. For an interesting account of the problem in this context,
      see Wilmsen, *Lindenmeier: A Pleistocene Hunting Society* (1976), pp. 45–58 and *passim*.

On the other hand, abandoning the reductionist red herring does not elim-
inate the problem of projection. Indeed, the correction of Winch's mistaken
view of natural science allows his argument about the impossibility of social
science to return with a vengeance. The demise of positivistic philosophy of
science has led to challenges to the notion of objective truth, or the idea that
there is a gradual convergence, over time, of theory to truth, in *natural* science. Some philosophers suggest that no sense can be given to the concept
of scientific objectivity. Rather, it is suggested that the conditions of verification of any theory are set by the particular 'community' of scientists who work
within the 'disciplinary matrix', 'paradigm', or '*Weltanschauung*' embodied in
it. Following this line of thought, the scientist-turned-sociologist Barry Barnes
compares the scientist to specialists in other types of knowledge, 'the prophet,
the astrologer, and the witch-doctor': '"true" like "good" is an institutionalized
label used in sifting belief or action according to socially established criteria'.[32]
Winch, too, begins from the fact that science is itself a social activity. As a result, 'to understand the activities of an individual scientific investigator we must
take account of two sets of relations: first, his relation to the phenomena which
he investigates; second, his relation to his fellow-scientists'.[33] The latter is as
determining as the former; it is only because they are taking part in the same
kind of activity, or cultural institution, and therefore accept the same criteria
of identity with respect to natural phenomena, that scientists communicate
with each other, cooperate in doing science, or, simply, do science at all. This is
important because, as Hayek noted in his attempt to distinguish natural from
social science, a given natural state of affairs will come under different descriptions if different criteria are applied to it.

Now, Winch argues, if sociology (or anthropology) is to be a science like the
natural sciences,

> The concepts and criteria according to which the sociologist judges that,
> in two situations, the same thing has happened, or the same action per-
> formed, must be understood *in relation to the rules governing sociological
> investigation*. But here we run against a difficulty: for whereas, in the case
> of the natural scientist we have to deal only with one set of rules, namely
> those governing the scientist's investigation itself, here what the sociolo-
> gist is studying, as well as his study of it, is a human activity and is there-
> fore carried on according to rules. And it is these rules, rather than those

---

32   Barnes, *Scientific Knowledge and Sociological Theory* (1974), pp. 66, 23.
33   Winch, *Idea* (1963), p. 84.

which govern the sociologist's investigation, which specify what is to count as 'doing the same kind of thing' in relation to that kind of activity.[34]

For example, the question whether two actions performed by members of a culture are both acts of prayer is a religious question: when a sociologist or anthropologist studies religion the criteria of identity of the institution studied must be taken not from sociology but from the religion studied itself. The anthropologist Marvin Harris has described the situation of the social scientist using Pike's distinction between 'emic' and 'etic' observation and analysis. The former

> have as their hallmark the elevation of the native informant to the status of ultimate judge of the adequacy of the observer's descriptions and analyses. The test of the adequacy of emic analyses is their ability to generate statements the native accepts as real, meaningful, or appropriate. In carrying out research in the emic mode, the observer attempts to acquire a knowledge of the categories and rules one must know in order to think and act as a native ... Etic operations have as their hallmark the elevation of observers to the status of ultimate judges of the categories and concepts used in descriptions and analyses. The test of the adequacy of etic accounts is simply their ability to generate scientifically productive theories about the causes of sociocultural differences and similarities ... Frequently, etic operations involve the measurement and juxtaposition of activities and events that native informants may find inappropriate or meaningless.[35]

Harris is careful to point out that the emic/etic distinction is not to be identified with a distinction between 'mental' and 'physical' phenomena, or with that between the realm of the meaningful and that of the observable. Indeed, according to Harris,

---

34 Ibid., p. 87.
35 Harris, *Cultural Materialism* (1979), p. 32. Harris's use of the emic/etic distinction is a subject of debate among anthropologists; see Anthony F.C. Wallace's review of *Cultural Materialism* in *American Anthropologist* (1980), pp. 423–6, and especially Fisher and Werner, 'Explaining Explanation: Tension in American Anthropology' (1978), pp. 194–218. The position taken in this paper is much closer to my own than is that of Harris. However, the passage I quote here from Harris does not commit the faults against which Fisher and Werner argue; since *Cultural Materialism* was published after their article, perhaps this indicates a responsive change in Harris's position. I am grateful for these references to Kenneth M. Kensinger of Bennington College.

The importance of Pike's distinction is that it leads to a clarification of the meaning of subjectivity and objectivity in the human sciences. To be objective is not to adopt an etic view: nor is it subjective to adopt an emic view. To be objective is to adopt the epistemological criteria ... by which science is demarcated from other ways of knowing.[36]

But – Winch would reply – there's the rub. A culture is *constituted* by a set of categories; nature has no point of view, but a way of life is defined by its practitioners'. The abandonment of Weber's dual method, of *Verstehen* plus 'objective' observation, necessary on Weber's own premises, apparently leaves no criteria available for judging the correctness of an observer's description or analysis of a culture, to the extent that this involves interpretation of the culture in categories foreign to it. The sin of projection seems an Original one for social science: even aside from the issues of 'subjectivity/objectivity' or of the difference between causes and reasons, the meaningful character of social phenomena seems to rule out the objectivity necessary if we are to be able to speak of a scientific study of society.

---

36   Harris, *Cultural Materialism* (1979), pp. 34–5.

CHAPTER 3

Understanding an Alien Society

Max Weber's example of explanatory 'adequacy on the level of meaning' –
explaining 'what is, according to our current norms of calculation or thinking, the correct solution to an arithmetical problem'[1] – sheds no light on the
problem of the interpretation of a culture different from the observer's own.
The action of arithmetical calculation can be interpreted with adequacy on the
level of meaning by reference to 'our' current norms because, it is assumed, the
calculator observed is obeying the same norms. But how is the behaviour of
members of a social group who believe (for example) in witchcraft or magic to
be interpreted by would-be scientific observers who do not share these beliefs?
The latter can state the means-end relation in the former's action ('They do
that to stop witches from attacking them') in terms of the group's own norms
of thought. But to the question, 'What is that man doing when he performs
his anti-witchcraft ritual?' the answer, 'Warding off witchcraft', is not very illu-
minating. Observers who are not of the group believing in witchcraft can place
this action in an explanatory framework only if they construct that framework
from motivational elements with which they are familiar. As Alfred Schutz
describes 'the paradox that dominates Weber's whole philosophy of social science', Weber

> postulates as the task of social science the discovery of intended mean-
> ing – indeed, the intended meaning of the actor. But this 'intended mean-
> ing' turns out to be a meaning which is given to the observer and not to
> the actor.[2]

But, objects Winch, this is to disregard the essence of culture – that it consists
of a set of shared meanings.

> When one is dealing with any kind of social 'things', ... their being ... social,
> as opposed to physical, in character depends entirely on belonging in a
> certain way to a system of ideas or mode of living. It is only by reference
> to the criteria governing that system of ideas or mode of life that they have

---

1 Weber, *Theory* (1947), p. 99.
2 Schutz, *The Phenomenology of the Social World* (1973), p. 234.

any existence as intellectual or social events. It follows that if the sociological investigator wants to regard them as social events ... he has to take
seriously the criteria which are applied for distinguishing 'different' kinds
of actions and identifying the 'same' kinds of actions within the way of life
he is studying. It is not open to him arbitrarily to impose his own standards from without. In so far as he does so, the events he is studying lose
altogether their character as *social* events.[3]

So far, Winch's position seems uncontroversial. Surely, however we intend to
analyse a set of cultural phenomena, we should strive to identify its elements
on the basis of the criteria of the culture studied. Examples Winch offers of his
recommended approach certainly seem to embody ordinary anthropological
common sense: e.g. that a psychoanalytical investigation of Trobriand neur-
oses 'could not just apply without further reflection the concepts developed by
Freud for situations arising in our own society', but would have to investigate
differences between (say) European and Trobriand concepts of family structure.[4] No psychoanalyst will take a prohibition of continuing investigation as
a methodological guideline, and no anthropologist or sociologist will admit to
imposing his own cultural categories 'arbitrarily' on the material of his field
research.

In fact, if put in these terms the purported contrast of social with natural
science is at least questionable. While nature (as I just said) can't be said to
have a point of view, it has been said to have or constitute an order of its own,
to which the community of scientists must accommodate its concepts even
while it assimilates natural phenomena to them by defining 'criteria of identity'. Physics is at any rate no more arbitrary than social science could be. So in
what way is Winch's argument an argument against the possibility of social science, rather than an argument against *bad* social science? To put it differently,
at what point, as Winch sees it, does the problem of projection reveal itself
to involve an irreducible conflict between the cultural point of view investigated and the categories of the observing scientific community? It might seem
as though Winch means to rule out analyses of a culture that go beyond what
members of it would give in explanation of their own behaviour. And this
would certainly rule out a great deal of what passes for social science. In general, as Fritz Machlup has pointed out, participants in a culture, even while they
may (and indeed must) know the rules and criteria regulating social behaviour

---

3   Winch, *Idea* (1963), p. 108.
4   Ibid., p. 90.

in that culture, may have only a very vague notion of how the parts of social life in which they participate fit together. Machlup offers the example of a Martian anthropologist – perhaps a colleague of MacIntyre's Martian – who observes the stock market and interviews its denizens.

> Since probably 999 out of 1000 persons working on the stock market do not really know what it does and how it does it, the most diligent observer-plus-interviewer would remain largely ignorant. Alas, economics cannot be learned either by watching or by interviewing the people engaged in economic activities. It takes a good deal of theorizing before one can grasp the complex interrelations in an economic system.[5]

This expresses in other words a point made in the previous chapter, that observation is meaningless except within categories provided by some theoretical orientation. In Winch's terms, science requires categories, and criteria for applying them, that go beyond observation, in the sense that they are created by groups of scientists. The workers in the stock market, lacking economic theory, are in a bad position to observe their own activity; the economist can do so just because he makes use of categories different from those of the insiders – categories of *economics*, rather than those of the mode of life that includes working in the stock market. But this means that he will employ criteria unused by the subjects of his study; he will impose his own standards in the analysis of their activity, though certainly (he will claim) not arbitrarily.

We can compare this case to that of a linguist's analysis of a language in terms of grammatical rules unelicitable from native speakers. Despite the view of some linguists that their theories represent not just theories of language but theories of language-users, explaining what the latter know as opposed to what they know how to do, it seems that at least we can say that linguists' knowledge of a language is different from a native speaker's, employing as it does theoretical concepts unknown to the native. Winch's views need not be taken to be in fundamental conflict with such a point of view. Indeed, in his eyes,

> the test of whether a man's actions are the application of a rule is not whether he can *formulate* it but whether it makes sense to distinguish between a right and a wrong way of doing things in connection with what he does. When that makes sense, then it must also make sense to say that

---

5   Machlup, 'If Matter Could Talk' (1969), p. 54.

he is applying a criterion in what he does even though he does not, and perhaps cannot, formulate that criterion.[6]

A closer look discovers difficulties in this formulation. It seems odd to say that someone is applying a criterion that he cannot formulate. As with the linguistic example, it seems to me more natural to speak of such a criterion as a feature of the theory or description of the man's behaviour, rather than of his knowledge of or obedience to a rule. And in any case Winch's formulation of the criterion for the use of 'criterion' threatens his own distinction between the researcher's criteria of identity and those of the object of study. To say that what is important is 'whether it makes sense' to formulate a criterion leaves open the question, to whom it must make sense, especially in the case where the actor himself 'cannot formulate the criterion'.

Winch's discussion of social rules, in fact, slides in a superficial way over the fact, a commonplace for ethnographers, that (in Malinowski's words) informants produce 'at best that lifeless body of laws, regulations, morals, and conventionalities which ought to be obeyed, but in reality are often evaded'.[7] The rules discerned by the investigator may therefore be quite different from those which may be elicited from an informant. As a result, Winch's test of when someone is following a rule permits ethnographers to formulate rules that contradict rules the natives studied would formulate. It is possible, however, to state Winch's criteria for non-projective social description in a way which avoids this problem. What is essential is not that all the categories employed by the outside researcher be given in the culture he studies, but that they do not contradict the classifications (not the rules) of social activity basic to the culture. In Winch's words,

> although the reflective student of society, or of a particular mode of social life, may find it necessary to use concepts which are not taken from the forms of activity which he is investigating, but which are taken rather from the context of his own investigation, still these technical concepts of his will imply a previous understanding of those other concepts which belong to the activities under consideration.[8]

---

6   Winch, *Idea* (1963), p. 58.
7   Malinowski, *The Sexual Life of Savages in North-West Melanesia* (1932), cited by MacIntyre, 'The Idea of a Social Science' (1973), p. 16.
8   Winch, *Idea* (1963), p. 89.

What kinds of social studies are unacceptable under Winch's rule? Most clearly, it rules out certain kinds of cross-cultural comparison of institutions and practices. Winch develops his argument to this point in the context of a critique (already noted) of Pareto's theory of 'residues' (psychologically determined constants of human behaviour) and 'derivations' (ideologies and institutions in which the residues are fleshed out). The example Winch discusses is Pareto's analysis of Christian baptism as a special case of the allegedly transcultural phenomenon of cleansing rituals. Pareto's idea is that this phenomenon is found in some form in all cultures; the various characteristics and explanations of the ritual in different cultures being derivations from it, abstraction from which reveals the essential, shared core (hence 'residue'). Christians share with members of other cultures a practice of ritual cleansing; what differ between them are the specific features of the practice and the reasonings explaining the efficacy of these various features.

Now, says Winch, 'a Christian would strenuously deny that the baptismal rites of his faith were really the same in character as the acts of a pagan sprinkling lustral water or letting sacrificial blood. Pareto, in maintaining the contrary, is inadvertently removing from his subject matter precisely that which gives them sociological interest: namely their internal connection with a way of living'.[9] It is what differentiates the two rituals that gives them sociological interest, that gives them, that is, the meanings they have for the members of the cultures concerned: for the connection of this element with the remainder of their ways of life is articulated exactly through the differences. To identify elements from different cultures is to commit the root error exemplified by Weber's behaviouristic description of monetary exchange (see Chapter 2, above). There the description of behaviour, divorced from the intended meanings of the actors, lost all sense as social description; here the sense given is a concoction by the social theorist rather than a representation of the actual social meaning involved. In Winch's words,

> ideas cannot be torn out of their context in that way: the relation between idea and context is an *internal* one. The idea gets its sense from the role it plays in the system. It is nonsensical to take several systems of ideas, find an element in each which can be expressed in the same verbal form, and then claim to have discovered an idea which is common to all the systems.[10]

---

9   Ibid., pp. 108–9.
10   Ibid., p. 107.

Pareto is no straw man here. It is worth remembering that E.E. Evans-Pritchard, in his attack on reductive theories of religion, cites Pareto approvingly as one who indicated an important direction for analysis and research: 'in spite of the superficiality and vulgarity and the confusion of his thoughts, Pareto saw the problem correctly'.[11] To take a contemporary example, Jack Goody, who has expended much effort in demonstrating errors of projection committed by anthropologists imposing their categories on the people they study, nonetheless describes the 'experimental method' of anthropology as 'comparing, not social systems or societies or cultures as such, but specified variables under different social conditions ...'.[12] Such comparison, of course, requires abstraction of the 'variables' from their social contexts, and identification of them between systems. And Marx certainly seems to fall, as Winch says he does, into the same class as Pareto with respect to Winch's critique, given the seeming similarity of the concept of 'derivation' to Marx's concept of 'ideology'.[13]

Winch's argument is in fact, as he wishes it to be, an attack on the sociological and anthropological tradition as a whole. The claim of the social sciences has always been that one can go beyond ethnological description to the formulation of generalisations applying across cultures, yielding knowledge fundamentally different from that possessed by cultural insiders. It is a commonplace that anthropology is 'a comparative discipline', both because 'by convention and established tradition of doing fieldwork in a culture other than one's own, the anthropologist uses a comparative framework in his study' and because anthropological theorising has always, 'with very few exceptions', involved 'intercultural or cross-cultural explicit comparison'.[14] Even in questioning the results of the anthropologists' search for cross-cultural regularities, Evans-Pritchard identified this search as the only path to a possible social science comparable to the physical sciences, concluding, 'my scepticism does not mean that I think that we should cease to look for such regularities as can be established by various forms of the comparative method'.[15]

---

11   Evans-Pritchard, *Theories of Primitive Religion* (1976), p. 98.
12   Goody, *Comparative Studies in Kinship* (1969), p. xiii.
13   See Winch, *Idea* (1963), p. 104.
14   Sarana, *The Methodology of Anthropological Comparisons* (1975), p. 15.
15   Evans-Pritchard, 'The Comparative Method in Social Anthropology' (1965), p. 35. The scepticism derives from his sense that social facts may well be 'so totally different from those studied by the inorganic and organic sciences that neither the comparative method nor any other is likely to lead to the formulation of generalizations comparable to the laws of those sciences. We have to deal with values, sentiments, purposes, will, reason, choice, as well as with historical circumstances' (p. 33).

It is obvious that some form of transcultural concept formation is essential to the project of a general science of society, and in this process social science imposes its own criteria on cultural phenomena. As Harris puts it, in his scientific tough-guy way,

> I cannot agree ... that our etic conceptual resources for the study of the behavior stream are dependent upon emic studies. The etic concepts appropriate for the study of the behavior stream are dependent on their status as predictive elements in a corpus of scientific theories.[16]

The very voice of the Anti-Winch! For Winch, to adopt such a perspective is to give up all hope of understanding social institutions or practices, to dignify the error of projection with the name of science. Gift exchange and monetary exchange may have in common the mutual transfer of objects between persons. But to treat them as in essence identical, as Mauss did in his celebrated essay on *The Gift*, ignoring the sense given to money by its place in the whole system of generalised commodity production, is to commit in theory an error of the magnitude of that committed in practice by the Indians who, without knowing it, sold the island of Manhattan.

Cultural elements can only be understood as rule-governed behaviour, and the rules exist only as stated in terms of the concepts of the particular culture concerned. To abstract from the specificity of these concepts is therefore to render cultural interpretation impossible. It follows that interpretation is possible only from *within* a culture, for rule-governed action can only be described and explained in terms of the relevant rules. '"Understanding", in social study, Winch says explicitly, 'is grasping the *point* or *meaning* of what is being done or said'.[17] So that

> If we are going to compare the social student to an engineer, we shall do better to compare him to an apprentice engineer who is studying what engineering – that is, the activity of engineering – is all about. His understanding of social phenomena is more like the engineer's understanding of his colleagues' activities than it is like the engineer's understanding of the mechanical systems which he studies.[18]

---

16   Harris, *Cultural Materialism* (1979), p. 41.
17   Winch, *Idea* (1963), p. 115.
18   Ibid., p. 88.

Social understanding, cultural interpretation, is, then, far from science; it is, for Winch, a species of Wittgensteinian philosophy: exploring the 'grammar' of a culture, it aims not to discover new 'facts' but to 'leave everything as it is'.

It will be instructive to compare Winch's position, as he himself has done, with that of Evans-Pritchard. In his book on *Witchcraft, Oracles, and Magic among the Azande* the anthropologist comes to ask,

> In writing about the beliefs of primitive peoples does it matter one way or the other whether one accords them validity or regards them as fallacious? Take witchcraft again. Does it make any difference whether one believes in it or not, or can one just describe how a people who believe in it, think and act about it, and how the belief affects relations between persons? I think it does make a difference, for if one does not think that the psychic assumptions on which witchcraft-beliefs are based are tenable, one has to account for what is common sense to others but is incomprehensible to oneself.[19]

If one believes there are witches, then a set of ideas about witches and how to ward off the effects of their evil powers will call for no particular explanation beyond normal perspicuity. If one does not share this belief, one will be moved to ask how others come to have it. What are the phenomena that give rise to this belief? Answering this question will involve thinking about witchcraft in a way quite different from that of the native believers. It may be placed into a transcultural classification of social practices which would seem shocking or absurd to its practitioners. For Evans-Pritchard, for instance, the belief in witchcraft he studied among the Azande was an example of 'ritual' behaviour, which he defined as 'behavior accounted for by mystical notions' – that is, by notions ascribing to phenomena properties that they do not have and that cannot be defended from the point of view of 'our body of scientific knowledge and logic'. The latter are 'the sole arbiters of what are mystical, common sense, and scientific notions' (although he adds that 'their judgements are never absolute').[20]

But Evans-Pritchard himself has acknowledged, as Winch points out, that the scientific thinking of Europeans just as much as the 'mystical' thinking of the Azande represents 'patterns of thought provided for us by the societies in which we live'.[21] According to Winch, it is such patterns of thought that define

---

19   Evans-Pritchard, *Witchcraft* (1976), p. 224.
20   Ibid., p. 229.
21   Cited in Winch, 'Understanding a Primitive Society' (1977), p. 80.

what is real and what is unreal. Since these patterns are represented by language, 'we could not in fact distinguish the real from the unreal without understanding the way this distinction operates in the language'. However, Winch holds, it is just on failing to do this that Evans-Pritchard's argument depends. The anthropologist wants

> to work with a conception of reality which is not determined by its actual use in language. He wants something against which that use itself can be appraised. But this is not possible; and no more possible in the case of scientific discourse than it is in any other. We may ask whether a particular scientific hypothesis agrees with reality and test this by observation and experiment ... But the general nature of the data revealed by the experiment can only be specified in terms of criteria built into the methods of experiment employed and these, in turn, make sense only to someone wholly conversant with the kind of scientific activity within which they are employed.[22]

The excellence of Evans-Pritchard's ethnology of the Azande has made this a standard case in the philosophy of the social sciences; I interrupt the discussion, therefore, for a brief summary of his findings. The Azande are (to use the ethnographic present) a people of the Sudan. Their lives are pervaded by interrelated notions of witchcraft and magic. Although Evans-Pritchard calls their beliefs 'mystical' in his sense, Zande do not experience the phenomena of witchcraft as unnatural, awe-inspiring, or mystical in our usual sense of that word. Rather they experience the activities of witches as a normal part of everyday life, as natural (although ill-understood) in their mode of operation. Witchcraft is held to be an inherited physical condition, which is activated by ill-feeling of various sorts. When activated it is responsible for damage to the interest of the person hated, envied, etc. If termites destroy a man's house, or his crops fail to grow, or he or his child becomes ill, the question immediately arises, who is responsible, via witchcraft, for the disaster. The witch responsible is discovered by the use of a number of oracles, with varying degrees of reliability. The most reliable Evans-Pritchard calls the poison oracle, as it consists of feeding a poison to chickens while asking questions, answered by the death or survival of the fowl. As the witch or witches responsible can only be individuals who bear ill will to the victim, only the names of such individuals are put to the oracle, which selects one, more, or none of them. This leads to a ritual in which

---

22   Ibid., p. 82.

the now-identified witch is asked to call off his powers, and to further testing to see if he has done so. If witchcraft results in death, vengeance is exacted by witchcraft. In addition, magical rites and medicines may be used for protection against witchcraft.

In describing all this, Evans-Pritchard makes a great effort to be faithful to the Zande point of view. Dealing with their conception of witchcraft as embodied in a physical substance, he goes so far as to tell us, 'I have only once seen witchcraft on its path', in the form of a bright light, without apparent explanation, passing in the night in the direction of a certain person's house. As it turned out, the next morning found an inmate of that house dead! In Evans-Pritchard's words, in Azande terms

> This event fully explained the light I had seen. I never discovered its real origin, which was possibly a handful of grass lit by someone on his way to defecate, but the coincidence of the direction along which the light moved and the subsequent death accorded well with Zande ideas.[23]

In the word 'real' here we feel working the point of view to which Winch objects as forcing the experience of the culture studied into the categorial framework of the investigator. Indeed Evans-Pritchard tells us that he himself 'always kept a supply of poison for the use of my household and neighbors and we regulated our affairs in accordance with the oracles' decisions. I may remark that I found this as satisfactory a method of running my home and affairs as any other I know of'.[24] At the same time, however, he states baldly that 'witches, as the Azande conceive them, clearly cannot exist'.[25] Given this, he asks a question no Zande would ask: why do people believe in them?

Could not the quarrel between the anthropologist and the philosopher be resolved by distinguishing explicitly between 'real-for-Azande' and 'real-for-Englishmen', with witchcraft falling under the former but not under the latter? After all, the starting point of anthropology is the fact of the differences between cultures. As Winch himself presents it, in an article taking issue with Evans-Pritchard along with some philosophical writers on the question of 'Understanding a Primitive Society', the aim of anthropology is to present an account of the beliefs and practices of a very different people 'that will somehow satisfy the criteria of rationality demanded by the culture whose concept

---

23   Evans-Pritchard, *Witchcraft* (1976), p. 11.
24   Ibid., p. 126.
25   Ibid., p. 18.

of rationality is deeply affected by the achievements and methods of the sciences, and one which treats such things as a belief in magic or the practice of consulting oracles as almost a paradigm of the irrational'.[26] The positive aspect of Evans-Pritchard's work is his attempt to sketch the rationality of the Azande, to communicate some sense of how the world can be experienced coherently through this particular pair of spectacles.

Evans-Pritchard's error, according to Winch, is that he does not stop there, but goes on to compare and contrast the Azande view of the world with his own along the dimensions of truth/falsity, reality/unreality, science/mysticism. But only by detaching it from its cultural context, its 'internal connection' with the rest of Zande life, can the sentence 'There are witches' uttered in Zande be equated semantically with the same sentence uttered in translation by an English anthropologist for whom it represents an untruth. This equation commits the sin of Pareto, identifying an element of one culture with an element of another, thus identifying them in terms that could not be accepted by members of the culture studied. 'Zande notions of witchcraft', Winch points out – in complete accordance with Evans-Pritchard – 'do not constitute a theoretical system in terms of which the Azande try to get a quasi-scientific understanding of the world'.[27] Therefore, the attempt to construe them as such is a mistake. But it is only in the context of such an attempt that it makes sense to discuss whether or not witchcraft is real-for-Englishmen, or whether it is true-for-Englishmen that there are witches. To compare 'real-for-Azande' and 'real-for-Englishmen' is to assume that 'real' itself means the same in both contexts. But this is exactly what Winch wishes to deny; since in his view 'to give an account of the meaning of a word is to describe how it is used'[28] and it is impossible 'to work with a conception of reality which is not determined by its actual use in language'.[29] Here we see that cultural relativism is leading to what we may call the incommensurability of languages: the view that there is simply no way in which an assertion in Zande about the reality of witches can be compared, with respect to truth-value, with an assertion in English about the reality of witches.

To go into more detail, it's not so much that the Azande have a different explanation than we do for some class of phenomena, but that they try to explain a class of phenomena that we do not recognise to exist. Thus, Evans-Pritchard tells us, 'Zande belief in witchcraft in no way contradicts empirical knowledge of cause and effect. The world known to the senses is just as real to

---

26   Winch, 'Understanding' (1977), pp. 78–9.
27   Ibid., p. 93.
28   Winch, *Idea* (1963), p. 123.
29   Winch, 'Understanding' (1977), p. 82.

them as it is to us.'[30] Mystical causation is not invoked as a substitute for natural causation, but as a supplement to it. If, for example, termites eat away the wood posts of a granary so that it collapses and kills people sitting beneath it, the only relationship we recognise between the fact that the building collapsed and the fact that these people were sitting beneath it is the accidental coincidence of two space-time paths at that place and moment.

> We have no explanation of why the two chains of causation intersected at a certain time and in a certain place, for there is no interdependence between them. Zande philosophy can supply the missing link. The Zande knows that the supports were undermined by termites and that people were sitting beneath the granary in order to escape the heat and glare of the sun. But he knows besides why these two events occurred at a precisely similar moment in time and space. It was due to the action of witchcraft.[31]

Similarly,

> we shall not understand Zande magic, and the differences between ritual behavior and empirical behavior in the lives of Azande, unless we realize that its main purpose is to combat other mystical powers rather than to produce changes favorable to man in the objective world.[32]

The point of magic, that is, is at least in part to influence the experienced course of events; but 'experience' here includes the experience of events as involving mystical phenomena. Indeed, such phenomena are part (to make a Winchian comment) of the 'objective' world of the Azande. The method of magic is not to affect the course of nature directly but to affect other mystical powers that are affecting nature.

What this reveals is that the Azande themselves make a distinction between ritual and empirical action. They do not use Evans-Pritchard's terms for this distinction, but they do distinguish between the categories distinguished by those terms: 'Azande undoubtedly perceive a difference between what we consider the working of nature on the one hand and the workings of magic and ghosts and witchcraft on the other hand, though in the absence of a formulated doctrine of natural law they do not, and cannot, express the difference as

---

30   Evans-Pritchard, *Witchcraft* (1976), p. 25.
31   Ibid., pp. 22–3.
32   Ibid., p. 199.

we express it'.[33] They speak, for instance, of witchcraft as the 'second spear' in an event, on the analogy of the second spear that may be thrown in the killing of an animal. 'Hence if a man is killed by an elephant Azande say that the elephant is the first spear and that witchcraft is the second spear and that together they killed the man'.[34]

All this leaves us with the problem of understanding the Zande assertion, 'There are witches'. When Evans-Pritchard tells us that 'Witches, as the Azande conceive them, clearly cannot exist', the subordinate clause embodies a claim to be contradicting the Zande assertion. And yet one sentence affirms the existence of witches from within the mystical worldview, while the other is obviously rejecting the existence of witches from the stance of a scientifically minded Englishman. Since the Azande themselves distinguish between empirical and mystical categories, Evans-Pritchard's view seems illegitimate. Perhaps a more proper statement would be something on the order of 'Witches cannot be said to exist in the same way as that in which we say that chairs and electrons exist'. As Winch makes this point, with respect to statements affirming the existence of the God of the Old Testament, what the reality of God amounts to

> can only be seen from the religious tradition in which the concept of God is used, and this use is very unlike the use of scientific concepts ... The point is that it is *within* the religious use of language that the conception of God's reality has its place, though, I repeat, this does not mean that it is at the mercy of what anyone cares to say; if this were so, God would have no reality.[35]

Thus one might say, the Azande and we agree, more or less, on how to think about the class of phenomena we call the realm of empirical causation. In addition, the Azande treat of a realm of phenomena whose existence we do not recognise. To say this is not to say that these phenomena don't exist, but to say that claims or denials of their existence are made in terms of categories which we do not use. For an English anthropologist to say, 'Witches do not exist' is for him to say that the system of concepts in which 'witch' has a central part plays no role in his culture. This is expressed very well by Evans-Pritchard himself:

---

33   Ibid., p. 31.
34   Ibid., pp. 25–6.
35   Winch, 'Understanding' (1977), p. 82.

I have often been asked whether, when I was among the Azande, I got to accept their ideas about witchcraft. This is a difficult question to answer. I suppose you can say I accepted them; I had no choice. In my own culture ... I rejected, and reject, Zande notions of witchcraft. In their culture ... I accepted them; in a kind of way, I believed them. Azande were talking about witchcraft daily, both among themselves and to me; any communication was well-nigh impossible unless one took witchcraft for granted.[36]

This is close to Winch's 'idea that the concepts used by primitive peoples can only be interpreted in the context of the way of life of those peoples'.[37] This statement taken by itself, however, is ambiguous. If taken to mean that analysis of a foreign culture involves stating the meanings of that culture's concepts in terms of their relationships to other portions of the culture, few would disagree. On the other hand, 'interpretation' in this context generally involves some notion of communication between cultures; it requires the idea of translation, the forging of a system of equivalences between the concepts of one culture and those of another. And this activity of translation is carried out by the anthropologist, on the basis of his cultural system. For this reason, translation is exactly what the Winchian is not interested in. As Winch's position is expressed by Sara Ruddick,

Understanding rather than translation is in question ... predicting and translating are not tantamount to understanding the sense they [the members of the alien culture] see in what they are doing ... Rather, as we realize in trying to understand others' or our own societies, making sense is a prolonged attempt to see the realities 'shown' in the concepts of others, i.e., to be able not only to say what they say, but to mean what they mean.[38]

This is related to Winch's view of reality as revealed differently in or through different cultural concept-systems; the attempt to establish equivalences between cultures is both an attempt to do something impossible and a waste of the opportunity to expand our experience of the world. Thus, according to Winch, in studying a culture S we are ideally

---

36   Evans-Pritchard, *Witchcraft* (1976), p. 244.
37   Winch, 'Understanding' (1977), p. 95.
38   Ruddick, 'Extreme Relativism' (1969), pp. 317–18.

seeking a way of looking at things which goes beyond our previous way in that it has in some way taken account of and incorporated the other way that members of *S* have of looking at things. Seriously to study another way of life is necessarily to seek to extend our own – not simply to bring the other way within the already existing boundaries of our own, because the point about the latter in their present form, is that they *ex hypothesi* exclude that other.[39]

Thus it is not that Zande concepts should be interpreted to match the distinctions we make. Rather, 'since it is we who want to understand the Zande category, it appears that the onus is on us to extend our understanding so as to make room for the Zande category, rather than to insist on seeing it in terms of our own ready-made distinction between science and non-science'.[40]

While this may be only polite, it is difficult to see how our categories can be expanded in this way, if Winch's conception of a culture is correct. Either we are dealing with two systems, each with its own criteria of rationality and truth, of which the elements are meaningless when dissociated from the whole. Or else it is possible to build conceptual bridges between elements of the systems, for us to understand a Zande category. But such bridges go in both directions: if we can stretch our categories to meet a Zande concept, the latter has *ipso facto* been brought into congruence with our ways of thought. 'I do not want to say', Winch writes,

that we are quite powerless to find ways of thinking in our society that will help us to see the Zande institution in a clearer light. I only think that the direction in which we should look is quite different from what [those I am criticising] suggest.[41]

Evans-Pritchard evaluates Azande magic as a semi-scientific procedure. But this is a mistake, says Winch. Compared to science, it is irrational, but only because this is the wrong dimension for comparison. Instead, Zande practice should be put on the dimension of religion, and Zande ritual compared to Christian prayer.

I do not say that Zande magical rites are at all like Christian prayers of supplication in the positive attitude to contingencies which they express.

---

39   Winch, 'Understanding' (1977), p. 99.
40   Ibid., p. 102.
41   Ibid., p. 103.

What I do suggest is that they are alike in that they do, or may, express an attitude to contingencies; one, that is, which expresses recognition that one's life is subject to contingencies, rather than an attempt to control these.[42]

It is ironic that Winch's argument has brought him here to do exactly what, in his book, he attacked Pareto for doing, when the Italian sociologist identified Christian baptism with Near-Eastern blood sacrifice. No evidence is given that the Azande would accept the identification of their magic with Christian prayer, even in the limited sense of Winch's comparison. And in fact, Evans-Pritchard's book shows clearly that the Azande regard magic as an attempt to *control* contingency. Checks for witchcraft are carried out before any major decisions are made or plans executed. Illness is combated by confrontation of the witch responsible and by magical healing.

Indeed, Winch ends up giving a standard anthropological-sociological explanation of ritual as a 'form of expression', or of 'drama', as 'ways of dealing (symbolically) with misfortunes'.[43] This is certainly not how the Azande themselves think of their behaviour; it is as much to put an interpretation drawn from our own culture on theirs as is the evaluation of witchcraft beliefs from the standpoint of natural science.

Winch has a retort available to this argument. Any comparison between cultures must of course involve reference to some common element. As we have seen, Winch rejects the possibility of the direct identification of elements of two different cultural systems. Comparison is possible, however, from this point of view, on the basis of some a priori-determinable set of 'prime' concepts applying to all societies. These concepts cannot, by the arguments discussed in the earlier part of this chapter, be discovered empirically (since any empirical comparison presupposes commensurability of elements from different cultures) but can be discovered through philosophical analysis. Such analysis, Winch claims, shows that

the very conception of human life reveals certain fundamental notions ... which ... in a sense determine the 'ethical space' within which the possibilities of good and evil in human life can be exercised ...: birth, death, and sexual relations ... The specific forms which these concepts take, the particular institutions in which they are expressed, vary very considerably

---

42 Ibid., pp. 104–5.
43 Ibid.

from one society to another; but their central position within a society's institutions is and must be a constant factor.[44]

It is hard to avoid seeing in this idea a close relative of Pareto's residues and derivations. All that distinguishes them in principle is the claim of a priori knowledge, derived from 'conceptual analysis'. But can this claim of Winch's be supported – even apart from general difficulties with the ideas of a priori cognition and conceptual analysis? Perhaps birth, copulation, and death are central concerns for all social systems, or even are bound to be. This can hardly be considered 'conceptually necessary', however; one can easily imagine a culture in which birth, for example, is not particularly signalised. In any case, the *notions* of birth, sex, and death don't seem to be the same for all cultures.[45] Similarly, when Winch speaks of 'the very conception of human life' he begs the question, *whose* conception? What his argument actually establishes is that he, Winch, is able to understand Azande witchcraft and magic by interpreting these practices and the conceptions they invoke as versions of his very different concepts of prayer, good and evil, and human life in general. It is almost anticlimactic to point out that in doing this he has totally departed from the sense that the Azande seem to give their own conceptions, since what seems to characterise these beliefs is an integration of ethical with empirical-manipulative practices; so that witchcraft is for them a matter both of physical action and of moral significance.[46] Winch objects to Evans-Pritchard's characterisation of Azande witch-talk as bad science, by saying that it is not science but religion, or 'ideas of good, evil, and the meaning of life'. But this separation of the religious/moral sphere from the scientific is quite ethnocentric. Evans-Pritchard's account makes it clear that for the Azande 'There are witches' expresses not just a moral or religious conception but also identifies certain persons as causal agents involved in certain unpleasant events.[47]

---

44   Ibid., p. 107.
45   An effective Winchian critique of just such an attempt to universalise the concept of 'kinship' can be found in Ruddick, 'Extreme Relativism' (1969), pp. 42 ff. It is hard to see why such an argument as she gives here would not apply equally to attempts at generalising categories like 'birth', etc.
46   See Evans-Pritchard, *Witchcraft* (1976), p. 53.
47   In Evans-Pritchard's words, 'For if the Azande cannot enunciate a theory of causation in terms acceptable to us, they describe happenings in an idiom that is explanatory. They are aware that it is particular circumstances of events in their relation to man, their harmfulness to a particular person, that constitutes evidence of witchcraft' (*Witchcraft* [1976], p. 24).

It seems, then, as though Winch's effort to find a standpoint outside of any culture, that would provide common terms of reference for the comparison of cultures, must be adjudged a failure. Morality and religion, along with birth, death, and sex, must be cast into the same wilderness of cultural relativism as rationality and truth. Indeed, how could it be otherwise? If concepts are defined by their role in cultural systems of concepts, how could there be a transculturally valid cultural analysis? We are thus squarely faced, as a consequence of Winch's relativism, with an inescapable incommensurability of cultures.

CHAPTER 4

Understanding and Explanation

The concept of incommensurability, to which Winch's views on the nature of social understanding brought us, is borrowed here from the philosophy of natural science. Winch himself suggests the analogy. Comparing magic to science, he writes, is 'like observing that both the Aristotelian and the Galilean systems of mechanics use a notion of force, and concluding that they therefore make use of the same notion'.[1] A purported translation of a certain Greek word by a certain (seventeenth-century) Italian word does not establish equivalence of meaning. Curiously enough, almost exactly this example has been employed by Paul Feyerabend, with T.S. Kuhn (at times) the best-known advocate of the incommensurability of scientific theories. In an article on the relations between successive theories of the same physical phenomenon, Feyerabend begins with the thesis that 'the meaning of a term is not an intrinsic property of it but is dependent upon the way in which the term has been incorporated into a theory' – its scientific use, to employ the Wittgensteinian catchphrase.[2] From this it follows that 'it is impossible to relate successive scientific theories in such a manner that the key terms they provide for the description of a domain D, where they overlap and are empirically adequate, either possess the same meanings or can at least be connected by empirical generalizations'.[3] They are literally incommensurable in the sense that measurement observations defined in terms of one theory T over the domain D will not be translatable into and so comparable with observations defined in terms of a theory T' covering the same domain.

> What happens when transition is made from a restricted theory T' to a wider theory T (which is capable of covering all the phenomena which have been covered by T') is something much more radical than incorporation of the *unchanged* theory T' into the wider context of T. What happens is rather a complete replacement of the ontology of T' by the ontology of T, and a corresponding change in the meanings of all descriptive terms of T' (providing these terms are still employed).[4]

---

1 Winch, *Idea* (1963), p. 107.
2 Feyerabend, 'Explanation, Reduction, and Empiricism' (1962), p. 68.
3 Ibid., p. 81.
4 Ibid., p. 59. For the example of 'force' in Aristotelian and Newtonian physics, see p. 57 and p. 89.

© KONINKLIJKE BRILL NV, LEIDEN, 2020 | DOI:10.1163/9789004414822_005

To complete the analogy with the anthropologist's position in the face of witchcraft, as Winch construes it, we note that Feyerabend explains that the incommensurability of theories 'does not mean that a person may not, on different occasions, use concepts which belong to different and incommensurable frameworks. The only thing that is forbidden for him is the use of both kinds of concepts *in the same argument*'.[5]

In the same way, as we have seen, on Winch's view Azande witchcraft and the scientific views current in England at Evans-Pritchard's time are metaphorically incommensurable.[6] There is no way in which the two worldviews might be jointly compared with reality, because (to recall Winch's words) 'both the distinction between the real and the unreal and the concept of agreement with reality themselves belong to' a given language or mode of life. This is not to say (Winch maintains) that there is no distinction to be drawn between something's being true and its being thought to be true. Thus 'God's reality is certainly independent of what any man may care to think, but what that reality amounts to can only be seen from the religious tradition in which the concept of God is used, and this use is very unlike the use of scientific concepts, say of theoretical entities'.[7] Truth, along with rationality, is relative not to an individual's thought but to the universe of discourse 'in terms of which an intelligible conception of reality and clear ways of deciding which beliefs are and are not in agreement with this reality can be discerned'.[8] Therefore, science cannot disprove witchcraft beliefs, although a scientific worldview can displace one in which witchcraft has a place, just as in Evans-Pritchard's own experience the language of witchcraft in certain contexts displaced his scientific worldview.

Despite Feyerabend's claim that such an extreme epistemological relativism constitutes an 'anarchist' theory of knowledge, to me the conservative implications linking the view that (in Feyerabend's phrase) 'anything goes' to the

---

5 Ibid., p. 83.
6 This is the situation pictured by the Seligmans in their study of the tribes of the Sudan: on the subject of magic, 'the black man and the white regard each other with amazement: each considers the behavior of the other incomprehensible, totally unrelated to everyday experience, and entirely disregarding the known laws of cause and effect' (Seligman, *Pagan Tribes of the Nilotic Sudan* [1932], p. 25; cited by Lukes, 'Some Problems about Rationality' [1977], p. 198).
7 Winch, 'Understanding' (1977), pp. 81–2. The last two assertions in this sentence are highly disputable. Alasdair MacIntyre has argued the contrary of the first on lines similar to those of my argument in 'Is Understanding Religion Compatible with Believing?'; Robin Horton has disputed the second in 'African Traditional Thought and Western Science', both in Wilson (ed.), *Rationality* (1977), pp. 62–77 and 131–71.
8 Winch, 'Understanding' (1977), p. 83.

'mainstream' social science Winch wishes to criticise are evident. Like Parsonian sociology, Winch's view does not lend itself to the explanation of change, either in ideas or in social reality, and contains an implicit justification of any social dogma in terms of the standards of truth and rationality set by the relevant mode of life. Perhaps this in part explains the popularity of Kuhn's earlier work, along with Feyerabend's, among political and social theorists, including some writers who claim descent from the Marxist tradition. Such views, I believe, indicate the abandonment of an earlier conception of human rationality as having the capacity to understand the world, including the problems and needs of human beings, and to reshape the social as well as the natural world the better to meet those needs.[9]

To say this is not to say that this sort of wholistic semantic relativism is wrong. But a sign that something is wrong with it is the fact that it is a member of that class of principles, of which the best-known member is the verification theory of meaningfulness, whose being true is incompatible with their being affirmed to be so. If the English can't understand the Azande, there is no way for Winch to grasp that the Azande have a different or any view of reality, and therefore no way for him to argue that the scientific and the magical worldviews are incommensurable! It seemed (pp. 37–38 above) that the anthropologist was stuck in the position of believing in witchcraft among the Azande, disbelieving it back in England, and striving not to let his two personae mingle and engage in cross-cultural conversation. But when we spell out Winch's view, we see that, if such a situation could occur, the anthropologist could not be aware of it, but would be condemned to the existence of a cultural Dr Jekyll and Mr Hyde. For it follows from the incommensurability of cultural systems that translation is not only insufficient but impossible. There is simply no way for the anthropologist to say meaningfully, 'The system of concepts in which "witch" has a central part plays no role in my culture'. This is because there is no way for the Englishman to know the meaning of the Zande words conventionally translated by 'witch' – *boro (ira) mangu* – as the rules of application of these words in Zande are quite different (on Winch's analysis) from those of

---

9 [If the idea of the incommensurability of scientific theories provided an accurate picture of the history of science, we would, of course, be forced to accept it despite its unpleasant political overtones. In fact, however, just as semantically homogeneous communities are to be found in science no more than in society generally, analysis of particular cases shows that the alleged incommensurabilities between alternative theories do not really exist (see, for example, the discussion of variant concepts of 'mass' in early twentieth-century physics of the electron in Galison, *Image and Logic* (1997), pp. 810 ff.). An interesting treatment of the issue is given by Beller, *Quantum Dialogue* (1999), Ch. 14.]

'witch' in English. Living in a scientific culture, we cannot even use the word 'witch' as the inhabitants of Salem, Massachusetts used it in the seventeenth century. There is no way in which we can say that the women executed there for witchcraft were not guilty – and furthermore no way in which we can say that they were. We can only say that in Salem at that time it was appropriate to use the words 'witch' and 'guilty' in the same sentence; but we can't say what these words mean.

This is the difficulty we encountered in Winch's self-contradictory attempt to extend the boundaries of one way of life to encompass another. Since (in Winch's words) 'the point about' our boundaries 'is that they *ex hypothesi* exclude that other', there is no way to know how to extend them or even what such an extension could mean. It is for this reason that Winch is forced at that stage of his argument to appeal to a transcultural a priori, the set of 'prime concepts' or 'fundamental notions' of human life. As we saw, however, this move is illegitimate. It depends on a non-relativist understanding of 'the very concept of human life' or the 'notions' of birth, death, and sexual relations.

The double bind that Winch has created for himself is actually visible in the statement of his relativistic account of truth itself. 'What is real and what is unreal shows itself in the sense that language has', he says.[10] But what is this 'real' that 'shows itself' in language? This way of speaking, which implies a reality *an sich*, independent of language, which as it were functions as a form of appearance for some 'real' content, is incompatible with Winch's view that 'the distinction between the real and the unreal and the concept of agreement with reality themselves belong to' a given language.[11] As Winch himself recognises, 'the idea that men's ideas and beliefs must be checkable by reference to something independent – some reality – is an important one. To abandon it is to plunge straight into an extreme Protagorean relativism, with all the paradoxes that involves'.[12] Therefore he wants to insist both that, within (e.g.) a religious language game, 'God's reality is certainly independent of what any man may care to think' and that 'what that reality amounts to can only be seen from the religious tradition in which the concept of God is used ...'.[13] It is evident that the victory over Protagorean paradox is only apparent: while the individual possesses in his tradition a standard of truth, he has no standard to judge the tradition itself. With respect to his religion, as opposed to a par-

10   Ibid., p. 82.
11   Ibid.
12   Ibid., p. 81; see also Ruddick, 'Extreme Relativism' (1969).
13   Winch, 'Understanding' (1977), pp. 81–2.

ticular religious perception, he is still in the position of Wittgenstein's private language-speaker, who buys several copies of the morning paper to assure himself that what it says is true.[14]

Winch argues that there is no culturally neutral notion of truth or rationality: that is the only conclusion consistent with his premises, despite the attempt to avoid their relativistic consequences. Science in particular, he claims, can lay no claim to objectivity, but is as much a matter of cultural subjectivity as witchcraft. But what, then, is the status of *these* statements – of Winch's philosophy? Winch offers no reason why we should accept his implicit claim that philosophy, alone among modes of thought, is not culturally relative, and that philosophers are more equipped than sociologists or anthropologists to win true knowledge of social life.

Winch remains trapped within the alternatives of 'objectivity' and 'subjectivity', which are only restated in the alternatives of a transcendentally established science and rationality, on the one hand, and relativism, on the other. Since it is the pair of alternatives that is the problem, neither provides a solution. For this reason, Winch's inconsistency, in trying to provide an interpretation of Azande witchcraft in terms of elements of his, English culture, is understandable. It is less his inconsistency than the principle he violates with it that is at fault. For if he were correct, and the gulf between cultures (as between the mental and physical pictures of the world) were so radical as to render them unintelligible except to insiders, then all communication between cultures would be impossible or else miraculous. And, difficult and laden with misunderstanding though it is, such communication is clearly a reality. We are as justified in asking how, not whether, anthropological knowledge is possible as Kant was in asking how natural science is possible. Winch himself knows perfectly well that it is possible to grasp a good deal of the workings of an alien culture, and this is why he tries to find parallels in his own society to make sense of that of the Azande.

Where has Winch gone wrong? One source of the problem can be located in a certain poverty of his categories for the analysis of cultural systems. Winch shares with radically relativist sociologists of knowledge a tendency to speak of 'thought', 'action', 'ideas', 'language', in a fairly undifferentiated way. This, for instance, allows Barry Barnes to speak of science as just one of numerous 'belief systems'. As G. Barden has shown, however, even Evans-Pritchard's account involves a more complex and subtle differentiation of the concepts of 'thought' and 'cultural system' than this allows for, suggesting that the analysis of cultural

---

14   See Wittgenstein, *Philosophical Investigations* (1953), Section 265.

categories should lead us to speak 'not generally about mentalities but about specific congeries of viewpoints, interests, and patterns of experience'.[15] Thus, it is not only that (in Evans-Pritchard's words) 'the Zande notion of witchcraft is incompatible with our ways of thought', but also that

> their intellectual concepts of it are weak and they know better what to do when attacked by it than how to explain it. Their response is action and not analysis. There is no elaborate and consistent representation of witchcraft that will account in detail for its workings ... The Zande actualizes these beliefs rather than intellectualizes them, and their tenets are expressed in socially controlled behavior rather than in doctrines. Hence the difficulty of discussing the subject of witchcraft with Azande, for their ideas are imprisoned in action and cannot be cited to explain and justify action.[16]

This is shown by the many cases Evans-Pritchard mentions in which Azande are totally undisturbed by contradictory elements in the doctrine of witchcraft. 'Azande do not perceive the contradiction as we perceive it because they have no theoretical interest in the subject, and those situations in which they express their beliefs in witchcraft do not force the problem upon them ... Azande are interested solely in the dynamics of witchcraft in particular situations'.[17] As Barden develops this,

> Azande fail to perceive a contradiction which springs at once to the mind of the alien European interpreter not because they are deficient in logic but because they have no theoretical interest in the subject; they do not work out the implications of a position because their interest is in action not in theory.[18]

In this way Evans-Pritchard distinguishes between theory, logically organised thinking that seeks to compare and systematise different views or aspects of a subject matter, and unsystematic thinking, which is tied to specific situations and actions, and does not seek for generalisation or systematisation.

As Barden points out, witchcraft doctrine, though classified by Evans-Pritchard as 'mystical', because of its (in his eyes) falsehood, falls for the Azande

---

15   Barden, 'Method and Meaning' (1972), p. 128.
16   Evans-Pritchard, *Witchcraft* (1976), pp. 31–2.
17   Ibid., p. 4.
18   Barden, 'Method and Meaning' (1972), p. 112.

under the heading of 'common sense' thinking which, in contrast to science's dependence on experiment and logic 'uses experience and rules of thumb'.[19] If we abstract for the moment from the question of truth and falsity, we can see that witchcraft shares this feature with non-mystical common sense, both among the Azande and in our own culture. The description of the Zande attitude to witchcraft quoted on the previous page fits, for instance, the attitude of most cooks, who are unable to account for their procedures and recipes in any systematic theoretical way. Another example is American patriotism; while many Americans think of themselves as 'patriotic', a conception expressed in a set of practices and attitudes, they are unable if asked to give a coherent account of this conception in the context of their social activity as a whole. As Evans-Pritchard puts it in a different context, 'It is not so much a question of primitive versus civilized mentality [he is criticising Levy-Bruhl] as the relation of two types of thought to each other in any society, ... a problem of levels of thought and experience'.[20]

We have discussed this distinction earlier in this book, in the form of the sort of sociological concept-formation of which Winch approves. The contrast between theoretical economics and the conception of their own activity that workers in the stock market may have is another case of the distinction I have been drawing here between common sense and theoretical thinking about social life. In formulating the principles of Zande witchcraft and magic, Evans-Pritchard was playing the role of the Martian investigator; in his words,

> I hope I am not expected to point out that the Zande cannot analyze his doctrines as I have done for him. It is no use saying to a Zande, 'Now tell me what you Azande think about witchcraft' because the subject is too general and indeterminate, both too vague and too immense, to be described concisely. But it is possible to extract the principles of their thought from

---

19   See Evans-Pritchard, *Witchcraft* (1976), p. 229.
20   Evans-Pritchard, *Theories* (1976), p. 91. It must be said that Evans-Pritchard's definitions of 'mystical', 'commonsense', and 'scientific' notions are not unproblematic. The first, for example, are defined as 'patterns of thought that attribute to phenomena supersensible qualities which, or part of which, are not derived from observation or cannot be logically inferred from it, and which they do not possess' (*Witchcraft* [1976], p. 229). Given what we know about the role of unobservables in science, and the far from straightforward connection of scientific concept-formation with observation, this definition would make it no easy task to discriminate between mystical notions and scientific errors.

dozens of situations in which witchcraft is called upon to explain happenings and from dozens of other situations in which failure is attributed to some other cause.[21]

Here, it is to be noticed, Evans-Pritchard has not overstepped the bounds of legitimacy set by Winch for social analysis. But once the various pieces of the witchcraft conception have been assembled into a system, even if the categories of this system are, in accordance with Winch's criterion, logically related to the categories of the culture itself, the way is open to the transcendence of those bounds. Barden makes the point very elegantly:

> The way in which common sense defines the world illuminates the notion of world view, world picture, etc. For the world defined by common sense is not necessarily formulated in other terms than recurrent actions. It is not necessarily objectified in words; neither is it reflected upon. It is lived. It may be reflected on subsequently, but first it is lived. The reflection may be performed by members of the community or by outsiders, but, whatever the case, reflection introduces the possibility of radical critique and it is at this point that the distinction between truth and error becomes relevant. For reflection takes its stand at the viewpoint of science or theory to enquire into the nature of common sense both as content and as performance.[22]

Why is it that 'reflection introduces the possibility of radical critique'? By 'reflection' Barden means the evaluation of a system of concepts from the viewpoint of science. Here one can imagine Winch objecting that we are once again witnessing simple cultural imperialism, with 'modern' culture set up as a judge on all others. But, as Barden points out, the contrast between Azande witchcraft and science 'is not a contrast between cultures but a contrast between patterns of experience or modes of meaning'.[23] What this means can be brought out by a look at a list, given by Evans-Pritchard, of factors supporting the Azande's belief in witchcraft. These factors fall into three categories. First, there is a

---

21   Evans-Pritchard, *Witchcraft* (1976), p. 23.
22   Barden, 'Method and Meaning' (1972), p. 116. To anticipate, compare Marx on the relation of social theory to social practice: 'Reflection on the forms of human life, hence also scientific analysis of these forms, takes a course directly opposite to their real development. Reflection begins *post festum*, and therefore with the results of the process of development ready to hand' (*Capital*, Vol. I, 1976 [1867], p. 168).
23   Barden, 'Method and Meaning' (1972), p. 117.

group which might be called sociological, such as 'political authority supports vengeance magic'. Second is a group of attributes characterising almost any intellectual system, such as the mutual support rendered by the different parts of the system and the general fact that 'a Zande is born into a culture with readymade patterns of belief which have the weight of tradition behind them'.

Both of these categories of factors hold for science as well as for Azande belief. It is this which gives force to assertions such as this of Barnes, that

> natural science should possess no special status in sociological theory, and its beliefs should cease to provide reference standards in the study of ideology or primitive thought. The sociology of science is no more than a typical special field within the sociology of culture generally.[24]

However, the picture changes when we look at the third group of factors. For instance, Evans-Pritchard tells us, 'Azande often observe that a [magical] medicine is unsuccessful, but they do not generalize their observations'. 'People do not pool their ritual experiences' for purposes of systematic evaluative comparison. 'Contradictions between their beliefs are not noticed by Azande because the beliefs are not all present at the same time but function in different situations'.

> The place occupied by the more important medicines in a sequence of events protects them from exposure as frauds. Magic is made against unknown witches, adulterers, and thieves. On the death of a man the poison oracle determines whether he died as a victim to the magic. If the oracles were first consulted to discover the criminal, and then magic were made against him, the magic would soon be seen to be unsuccessful.[25]

As a result, the formal properties constituting logical and methodological flaws are not apparent to the Azande.

This observation gives an ironic resonance to Winch's insistence that 'ideas cannot be torn out of their context' since 'the idea gets its sense from the role it plays in the system'. This position appears plausible only so long as we forget that concepts can play different roles in systems, and that there are different sorts of systems of categories. A given idea, that is, can be situated in more than one context. Thus even if we set aside the question of the truth or fals-

---

24 Barnes, *Scientific Knowledge* (1974), p. 43. See Winch, 'Understanding' (1977), p. 89.
25 Evans-Pritchard, *Witchcraft* (1976), pp. 202–4.

ity of Azande witchcraft beliefs in order to ask merely how they are related to other elements of Azande life, we will be drawn back to the matter of truth in order to answer this question. Scepticism is not presupposed by such an inquiry, for one can certainly attempt a sociological study of the rise of scientific theorising in the last few hundred years without believing that science is all mystification. The fact that natural science has made possible an expansion of human beings' conscious interaction with nature, thanks to its ability to provide reliable information about the world, will, however, play a role in the explanation of the relation of, say, physics to the rest of modern social life. Similarly, the unsystematic, illogical, unscientific aspect of Azande witchcraft doctrine, its unsuitability for providing such information and in fact its failure to do so, are data for the investigation of 'the role it plays in the system'. The difference between such different modes of thought is not gainsaid by the recognition that concept-formation in both science and magic is not a matter of transcribing into ideas data supplied by some conceptually neutral experience, but one involving individual and social imagination. In Barden's words,

> the constructs of imagination are not either true or false; they are created images and as such go beyond sense experience ... But the imaginative complex, when it permeates a people's entire world, is necessarily linked with the operative desire for truth and rests on the popular assumption that it is true. The truth or falsity of the position is important precisely because of the implicit procedure and not to be able to make the distinction is to fail ultimately to rise above description. This implies that the scientific grasp of another position is not identical with an understanding of the position available to the original actors. It is to know how they know, and what they know, but not as they know.[26]

The problem we have been examining here has been discussed by J.L. Mackie under the rubric of 'ideological explanation'. He defines an ideology, in a manner clearly derived from Marx, as

> a system of constructs, beliefs, and values which is characteristic of some social class (or perhaps of some other social group, perhaps even of a whole society) and in terms of which the members of that class (etc.) see and understand their position in and relation to their social environment and the world as a whole, and explain, evaluate, and justify their actions,

---

26   Barden, 'Method and Meaning' (1972), p. 126.

and especially the activities and policies characteristic of their class (etc.) ... At least some of the beliefs and concepts in the system are false, distorted, or slanted, and at least some of the activities sustained and guided by that ideology have a real function differing from that which, in the ideology, they are seen as having.[27]

Zande witchcraft can be considered an ideology under this definition, which has obvious similarities to Evans-Pritchard's definition of 'mystical' belief. It might be thought preferable to omit the reference to a putative 'real function' of ideology, given the difficulties long recognised in functionalist theory. I have left it in, because it indicates the need for some explanation of the maintenance of false beliefs, which must play a role in social life different from the one they are thought to play by people who believe them to be true.

In terms of his definition Mackie distinguishes two senses of 'ideological explanation': (1) 'an explanation given within and in terms of an ideology by some of those whose ideology it is'; and (2) an explanation, given by someone who does not share the ideology, either of the ideology itself or of actions as resulting from people's holding the ideology. In short, the distinction is that between what we have called the insider's and the outsider's views of a culture; in Harris's terminology we could refer to emic and etic explanation. I will illustrate these modes of explanation with a few aspects of Zande witchcraft for which explanations might be asked.

1. Why, among the Zande, is a suspected witch often approached in a conciliatory way rather than with open anger?
2. Why are men not accused of bewitching their wives?
3. Why do the Zande believe in the phenomenon of witchcraft?

Ad 1. The Zande answer to the first question is that since witchcraft is activated by ill-feeling on the part of the person in whose body witchcraft-substance resides, the last thing a victim ought to do is provoke more ill-feeling. An anthropologist, noting this in his ethnography of the custom, would probably give the same explanation. The question concerns the inner articulation of the concept of witchcraft, and its answer must be stated in terms of that concept, which is the same for outsider and insider alike.

Ad 2. Zande explain that men don't practice witchcraft against their wives because 'no man would do such a thing as no one wishes to kill his wife or cause her sickness since he would himself be the chief loser'.[28] The anthro-

---

27 Mackie, 'Ideological Explanation' (1975), p. 184.
28 Evans-Pritchard, *Witchcraft* (1976), p. 9.

pologist M.D. McLeod, in contrast, explains this fact as a case of the general principle that accusations of witchcraft are 'confined to oracle owners' in a situation in which 'control of women and control of oracles [are] in the hands of roughly the same people'.[29] Women, that is, are in the first place not in a position to make authoritative accusations against their husbands. Secondly, McLeod argues that the practice of witchcraft accusation as a whole is part of a system of competition between men and kin-groups of men for social power, which is to a large extent embodied in power over women. The first part of this explanation does not contradict the native's explanation; so far it seems compatible with Winch's criterion of propriety in anthropological accounts. But the second part of the explanation shows that it is not really in accord with Winch, because it involves an account of the witchcraft system as a whole which, if comprehensible to a Zande, would be unacceptable to him or her, since by explaining the distribution of accusations as a function of social power relations it conflicts with native explanation in terms of the empirical reality of witchcraft. Implicit in McLeod's discussion is the assumption that the correlation of power relations and witchcraft beliefs is not coincidental, and that in some manner, to be determined by anthropological theory, the former explains the latter.

Ad 3. For the Zande the answer to the third question is simple: they believe in witchcraft because the evidence for its reality is all around them. Evans-Pritchard's explanation has already been quoted in part:

> Witches, as the Azande conceive them, clearly cannot exist. None the less, the concept of witchcraft provides them with a natural philosophy by which the relations between men and unfortunate events are explained and a ready and stereotyped means of reacting to such events. Witchcraft ideas also embrace a system of values which regulate human conduct.[30]

For example, Evans-Pritchard analyses the explanation of death as caused by witchcraft by pointing out that

> death is not only a natural fact but also a social fact. It is not simply that the heart ceases to beat ... but it is also the destruction of the member

---

29 McLeod, 'Oracles and Accusations among the Azande' (1972), pp. 170, 176; the point had already been made by Evans-Pritchard: 'The poison oracle is a male prerogative and is one of the principal mechanisms of male control and an expression of sex antagonism' (*Witchcraft* [1976], p. 130).

30 Evans-Pritchard, *Witchcraft* (1976), p. 18.

> of a family and kin, of a community and tribe ... Among the causes of
> death, witchcraft is the only one that has significance for social behavior.
> The attribution of misfortune to witchcraft ... gives to social events their
> moral value.[31]

As with the previous example, this explanation depends on implicit reference
to an anthropological theory – to a particular conception of culture and of the
role and mechanism of symbolisation in social life. Whatever we may think
of this theory, this example serves to make clear the logic of the outsider's
explanation. Given that the native's own explanation ('Witches exist') cannot
be true, how are we to explain his belief? And it is necessarily explained in
terms foreign to the native's view of the world, in terms derived from the culture, general and scientific, of the observer. The meaning of witchcraft-belief
which is being explicated is not the meaning for Azande, but the meaning for
the anthropologist. After all, the explanation here provided isn't one that the
native even feels any need for. The meaning here assigned 'witchcraft' is not
the semantic meaning of the term within Zande culture, but spells out the relation between the idea of witchcraft as a social fact and other facts about Zande
society.

Ideological explanation of the second type appears, in Mackie's words,

> radically faulty because it explains something other than what it purports
> to explain. It purports to explain actions and social behavior: but the ideo-
> logy of the relevant class (society, etc.) is a constituent of those actions
> and behavior – they are essentially however the agents see them to be –
> so any explanation from outside, which treats the ideology merely as a
> fact, causally related to actions and behavior described in other terms,
> will invariably neglect something essential.[32]

As I have already suggested, the error in this line of reasoning lies in the slide
from the fact that ideology is a constituent of actions and behaviour to the
assertion that the latter are therefore 'essentially however the agents see them
to be'. And as we saw, there is a sense in which this is true, as a result of the
fact that human life is organised by culture. Accounts of pre-capitalist goods
transfers that describe these actions as though they were market transactions
will fail to be accurate social descriptions; ethnography of the Azande begins

---

31   Ibid., p. 25.
32   Mackie, 'Ideological Explanation' (1975), p. 187.

with the description of certain practices as involved with witchcraft. But it is also true that the behaviour of social actors (and here is Mackie's correction of the view described above)

> has some partial causes of which they are unaware, that it will have effects that are not included in their purposes, and indeed that it may have an unknown function in the sense that these unknown partial causes include the fact that such actions tend to produce these effects which were no part of the agents' purposes. Thus there is a sense in which the agents can be wrong about what they are doing.[33]

Azande engaged in warding off witchcraft are (apart from the 'placebo effect') not really curing illness or ensuring successful hunting, although this is the (ideological) description they would give of their activity. The way is at least open to ask what else they might be doing, without being aware of it, and this question even must be posed once we recognise that 'witches, as the Azande conceive them, clearly cannot exist'. In any case,

> a full description of what is going on must and can take account of both aspects, both of how their actions appear to the agents and of what is not apparent to them: it can recognize distortions as part of what there is without itself becoming distorted.[34]

Mackie's account takes us far beyond the bounds set by Winch's criterion of faithful interpretation, because it places concepts clearly in their real context, that of action. This is a crucial point: that action is a context, and not an equivalent, for concepts and language; or rather that action is a more general term, covering thought and language as well as other forms of action. For the shallowness of Winch's theory derives in part from his simple identification of concepts and behaviour. For Winch, actions embody concepts; if they speak louder than words it is because they are like words, forms for conceptual contents. The context from which ideas cannot be torn is for Winch a 'system of ideas'. For instance, Winch tells us that 'social relations between men exist only in and through their ideas' or again that 'our language and our social relations are just two different sides of the same coin. To give an account of the meaning of a word is to describe the social intercourse into which it enters'.[35]

---

33   Ibid., pp. 193–4.
34   Ibid., p. 194.
35   Winch, *Idea* (1963), p. 123.

This is why 'the relation between idea and context is an internal one', to be understood by conceptual analysis.[36] Understanding a society is, on this view, nothing other than grasping the meanings of ideas in their systematic inter-connection.

To take a concrete example, 'war' for Winch is not a concept devised to sig-nify a type of social behaviour; rather it is the use of this concept which makes the behaviour the kind of behaviour it is. These are not, however, the only alternatives. True, war cannot be characterised as a kind of 'behaviour' but only as a cultural institution. But it does not follow from this that the institution is correctly understood by the natives whose institution it is, nor that actions always correspond to ideologies, so that acts of war always have the character claimed for them by actors.

To begin with, as the Azande witchcraft example brings out, to claim this is ultimately to agree with Hayek that 'any knowledge which we may happen to possess about the true nature of the material thing, but which the people whose action we want to explain do not possess, is as little relevant to the explanation of their actions as our private disbelief in the efficacy of a magic charm will help us to understand the behavior of the savage who believes in it'. But we have already seen that an unbeliever's account of magical prac-tices will be quite different from a believer's. Whatever the linguistic reality of witchcraft, it either does or doesn't cause crops to fail, make for bad hunt-ing, or cause children to die from fevers; and deciding this question may be a matter of life or death for the people involved. The persistence of belief in methods that don't work calls for some explanation, even though we normally can't expect those for whom those methods embody basic tradition to heed that call.

There is also a further, and most important, point, which I have already hin-ted at. Social intercourse is richer than language. It comprises a vast range of objects and practices which may lie outside the field of verbal representation or even contradict it. Even in those areas of social action which are organised through language, if we identify the meaning of a word or idea with the social behaviour in which it plays a role, it may easily turn out that meaning in this sense will differ from the meanings elicitable from natives. How do we, then, 'describe the social intercourse' in which a concept has its place? If we do it in the language of the natives, we have learned nothing; if we do it in our the-oretical language, we will describe social action differently from a native and perhaps in a way a native might not recognise.

---

36   Ibid., p. 107.

Winch's view assumes a double coherence of worldview: both internal coherence and coherence in its relation to action. As Alasdair MacIntyre has pointed out, however,

> The criteria implicit in the practice of a society or of a mode of social life are not necessarily coherent; their application to problems set within that social mode does not always yield one clear and unambiguous answer. When this is the case people start questioning their own criteria. They try to criticize the standards of intelligibility and rationality which they have held hitherto. On Winch's view it is difficult to see what this could mean.[37]

This is unrealistically optimistic: actually it takes a good deal more than conceptual ambiguity or incoherence to spur people to criticism of their conceptual schemes. These features have perhaps even been typical of such schemes; the main point is that, whether or not people are questioning their own criteria, the latter may well be open to question.

Ernest Gellner provides a complex example in his description of the institution of the *agurram* of Moroccan Berber society. This is a man who is held to be selected by God to be a spiritual leader, adjudicator of inter-group disputes, etc. Selection is indicated by endowment with magical powers, as well as by the combination of generosity, prosperity, pacifism, unconcern with material wealth, and other traits. Winch might take exception to Gellner's statement that 'in reality' election of an *agurram* is not by God but by his fellow tribesmen, as to Evans-Pritchard's statement of his disbelief in witchcraft. But he must admit that an *agurram* cannot really have at once all the characteristics with which he is credited and on which his identification as an *agurram* depends. Lack of concern with material wealth in reality spells the end of prosperity and the possibility of generosity. Were this internal contradiction to be recognised, the social category would disintegrate, just as it would were the reality of non-divine election admitted, since this would rule out the neutrality within the tribal system essential to the social role of the *agurram*.[38]

This brings out another flaw in Winch's identification of social concept and social action: its neglect of the fact that 'concepts generally contain *justifications* of practices, and hence that one misinterprets them grossly if one treats them simply as these practices, and their context, in another dress'.[39] As Gellner concludes,

---

37   MacIntyre, 'Is Understanding Religion Compatible with Believing?' (1977), pp. 67–8.
38   See Gellner, 'Concepts and Society' (1977), p. 43.
39   Ibid., pp. 44–5.

One might sum all this up by saying that nothing is more false than the claim that, for a given assertion, *its use is its meaning*. On the contrary, its use may depend on its lack of meaning, its ambiguity, its possession of wholly different and incompatible meanings in quite different contexts, *and* on the fact that, at the same time, it as it were emits the impression of possessing a consistent meaning throughout – on retaining, for instance, the aura of a justification valid only in one context when used in quite another.[40]

For example, the social use of the concept of 'nobility' in Western Europe depended on the ambiguity between the identification of certain highly-valued behaviour traits – courage, generosity, loyalty, etc. – and the identification of a certain social position, which allowed people in that social position *ipso facto* to lay claim to the traits. As we will see in the next chapter, Marx tries to bring out similar features of the ideology of economics, attempting to show the role of the essential ambiguity of key concepts in the continuing existence of the current social system. (Thus 'capital' for David Ricardo – as for subsequent economic theory – signifies at once raw materials, tools, and other means of production, and money invested with the aim of realising a profit.)

To say all this is, in a way, to restate the position of Winch with which we began this chapter, only to draw from it consequences very different from his. Winch is quite right to emphasise that scientific understanding of a culture involves the interpretation of the meaning of its elements in terms other than those of the culture itself. It is in his attempt to prove the impossibility of such interpretation that Winch involves himself in his uncomfortable relativism.

Winch is, I think, correct in contrasting translation with understanding. Gellner is like many in describing the anthropologist's problem as one of translation:

> The situation facing a social anthropologist who wishes to interpret a concept, assertion, or doctrine in an alien culture, is basically simple. He is, say, faced with an assertion S in the local language. He has at his disposal the large or infinite set of possible sentences in his own language. His task is to locate the nearest equivalent or equivalents of S in his own language.[41]

---

40   Ibid., p. 45.
41   Ibid., p. 24.

Now, clearly, this task cannot be accomplished without an understanding of enough of the culture to pick a justifiable equivalent for *S* in the home language. But – and this is the dilemma Winch's argument brings out – understanding seems to require translation just as translation requires understanding. With this dilemma Winch is stuck: the cultural relativity of concepts spells their incommensurability.

The origin of this problem lies in construal of the problem of interpretation as one of preservation of meaning in the first place. Construed in these terms, the problem is insoluble, since an outsider cannot share the insider's experience of his worldview and so the meaning of its conceptual constituents for him. The hermeneutic circle will be unbroken. A solution appears, however, as soon as we look at translation as both based on and enabling a reconstruction of the culture in its non-linguistic and non-conceptual aspects – what I referred to above as the context of action. This, as suggested by the Wallace article cited in Chapter 2, is what goes on in real cases of radical, and non-radical, translation. If we ask in what form translation is given,

> It may seem that the answer Quine and Davidson give, that translation is a recursive correlation of expressions with expressions, and ultimately of sentences with sentences, is obviously correct, even tautological. There may be a sense of 'translation' in which their answer is tautological and true. But if we mean by the form of the translation the form in which the meaning of the [Linear B] tablets is explained to us, their answer is wrong. The real explanations have the form of a translation in the Quine-Davidson sense plus a gloss ... The scholars in this business never give the meaning of a tablet simply by a word for word translation: there is always a gloss. And when you think of it, in ordinary life there is almost always a gloss when we say what someone meant or said or thought.[42]

The gloss, of course, refers to the non-linguistic context; since it is in the translator's language, and is not part of the translated material, it depends upon the making of cross-cultural comparisons as a means to knowledge of the alien culture.

My intention here has not really been to re-enter the problem of translation, but rather to stress my conclusion that translation, or more generally the preservation of meaning between cultures, is not the prime operation of social study. The anthropologist may even find a meaning for cultural features that the

---

42   Wallace, 'Translation Theories' (1979), pp. 126–7.

natives do not understand. To say this is not to contrast 'social-conceptual' with 'physical' description. As we saw in Chapter 2, social action is itself observable; it is open to alternative descriptions. One may therefore dispute a social actor's account of his own behaviour. But this does not mean that a social scientific analysis of witchcraft ritual in terms of power relations among the Azande is a description of a different realm of phenomena than the native's own description in terms of witchcraft. This is shown by the fact that the anthropologist's analysis employs the native's concepts, although not only them. Both accounts, the insider's and the outsider's, are descriptions and explanations of the same actions. These actions will not then have the same meaning for the outsider as for the insider, but they will not therefore become meaningless, because they will not have lost their connection with social life as a whole. Possibly, it is only in the anthropological account that that connection will be revealed; in any case, it will be spelled out in terms different from those of the natives. This does not mean that the natives' experience of their own culture becomes of no account. Just the opposite: it is among the phenomena with which social theory must grapple. It is not by its ability to substitute itself for a culture that a social theory is to be judged but by its ability to explain it.

CHAPTER 5

Foundation and Superstructure

Recognising that culture provides the structure of social action, I gave reasons in Chapter 2 for abandoning the use of 'subjectivity' to single out the character of social facts, because of the misleading nature of the contrast with 'objectivity'. Similarly, in the last chapter I argued for dropping 'meaning' as the social scientist's object of study. The paradoxes of Weberian sociology and its variants, as well as of Winch's philosophical approach, turn out to arise at least in part from the difficulty of reconciling the meanings (for insiders) of cultural elements with the demand for (outsider's) scientific truth – an opposition again conceived of as one of 'subjectivity' to 'the objective'.

An excellent epitome of the position I have been opposing is provided by a recent historical article on European witchcraft. Basing himself on Winch's work, Stuart Clark argues that Renaissance demonology is to be understood in terms of the total worldview within which it had its place, and that

> part of what we mean when we speak of a 'world-view' at all is surely that its constituents need no other explanation than their coherence with one another. The primary characteristic of demonological texts as historical evidence is not their supposed[!] unverifiability but their relationship to what J.L. Austin called a 'total speech situation': their meaning for the historian may be thought of as exactly symmetrical with their original meaning as linguistic performances.[1]

Clark admits the cost of this preservation-through-reconstruction of meaning: 'A contextual reading of Renaissance demonology may not help us answer the major questions about the genesis or decline of the European "witch craze" ...'.[2] But presumably what historians, along with anthropologists and social theorists generally, are after *is* the explanation of social phenomena, including those of meaning. Ideological explanation in the first of Mackie's two senses – explanations given in the terms of a given ideology – is not enough; we want also explanation in his second sense, explanation *of* the ideology. Exactly because the conceptual aspect of culture has a systematic character, though as we saw it

---

1 Clark, 'Inversion, Misrule, and the Meaning of Witchcraft' (1980), p. 127.
2 Ibid., pp. 126–7.

is not as consistent as Winch imagines, satisfying the demand for explanation of this type will involve relating concepts to non-conceptual elements of social life, since otherwise we will remain within the circle of concepts.

Clark objects to 'the explanation of ... witchcraft beliefs in terms of social and socio-psychological determinants'[3] on the ground that this bypasses meaning by reducing these beliefs to the status of epiphenomena of non-ideological social factors. If we are to make some sense of the concept of ideological explanation we must recognise the justice in this point of view, despite its ultimate untenability. To begin with, we may reject explanations of (e.g.) witchcraft that claim that witch-talk is 'really' talk about social relationships, morality, etc. Such explanations remain on the terrain of 'meaning' and on this terrain they fail exactly as does the similar effort to explain moral terms as 'really' expressions of feeling or descriptions of social conventions. 'Murder is wrong' does not *mean* 'I don't like murder' and 'This is the work of witches' spoken among the Azande does not *mean* 'There is bad will at work in this or that social relationship'. As Martin Hollis expresses this,

> A Zande, who believes he has been bewitched, surely does not believe that he has offended some social authority or other. If he did, he could perfectly well say so. He surely believes, rather, that he is the victim of supernatural interference.[4]

Similar objections can be made to a classical schema of ideological explanation, that which operates in terms of social 'function'. Clifford Geertz describes this schema wittily:

> A pattern of behavior shaped by a certain set of forces turns out, by a plausible but nevertheless mysterious coincidence, to serve ends but tenuously related to those forces. A group of primitives sets out, in all honesty, to pray for rain and ends by strengthening its social solidarity ... The concept of latent function is usually invoked to paper over this anomalous state of affairs, but it rather names the phenomenon ... than explains it.[5]

This remark holds true of both the chief forms of anthropological functionalism. To begin with, the idea that items in a culture are to be explained

---

3   Ibid., p. 98.
4   Hollis, 'Reason and Ritual' (1973), p. 38.
5   Geertz, 'Ideology as a Cultural System' (1964), p. 56.

by demonstrating the interactions between them (e.g. the correspondence between a hierarchy of gods and a social hierarchy) does not take us very far. If we ask to what questions such functional analyses can supply answers, we find only questions of the form, 'Is there a pattern of interaction between the various elements of a culture?' This is a question of some interest. The origin of functionalist theory lay in the rejection of speculative historical explanations of cultural traits. Against this functionalism insisted, first, that traits could not be understood in abstraction from their systemic context, and second, that explanations had to be given in terms of real information, gathered in fieldwork, and not in the form of guesses, more or less inspired, about (unknown) earlier periods. But this form of explanation cannot supply answers to questions about the origin and development of the culture as a whole.

These limits of this form of functionalism are avoided by a second version, in which social traits are to be explained by their causal efficacy in satisfying certain biological-psychological needs (Malinowski) or in maintaining the existence of the social system (Radcliffe-Brown).[6] Aside from problems peculiar to each of these patterns of explanation, both suffer from the flaw that (in Maurice Mandelbaum's words) such 'generalizations permitted no deductive consequences with respect to the specific nature of the practices of the peoples which they investigated'.[7]

It is interesting to note that Marx made a similar point in criticising Ludwig Feuerbach's speculative anthropological study of Christianity. Feuerbach believed that his book, *The Essence of Christianity* (1841), demonstrated that the Christian idea of God represented a hypostatisation of the social nature of the human species. This he argued for by showing that the attributes of God are the attributes of humanity. The content of the argument, despite its philosophical form, is very much that of the symbolic interpretation of culture. Against this mode of analysis Marx commented that 'it is ... much easier to discover by analysis the earthly kernel of the misty creations of religion than to do the opposite, i.e. to develop from the actual, given relations of [social] life the forms in which these have been apotheosized'.[8] To do this would be to meet Mandelbaum's

---

6 [See Malinowski, *Sex and Repression in Savage Society* (1927); Radcliffe-Brown, *Structure and Function in Primitive Society* (1952).]

7 Mandelbaum, 'Functionalism in Social Anthropology' (1969), p. 324. To my taste, Mandelbaum's phrase suggests an unwarranted attachment to the covering-law model of scientific explanation. His point can be extended, however, since functionalist generalisations also permit no theoretically warranted consequences with respect to specific cultural phenomena.

8 Marx, *Capital*, Vol. I (1976 [1867]), p. 494.

challenge to functionalism: to explain the particular categories of a culture on the basis of their relation to others of its features. How did Marx think this can be accomplished?

The following passage, from Marx's and Engels's early work against the 'German ideology', again has Feuerbach in view:

> In direct contrast to German philosophy which descends from heaven to earth, here [in their own work] it is a matter of ascending from earth to heaven. That is to say, not of setting out from what men say, imagine, conceive ... in order to arrive at men in the flesh; but setting out from real, active men, and on the basis of their real life-process demonstrating the development of the ideological reflexes and echoes of this life-process.[9]

In the discussion of Winch we saw the inadequacy of forming an image of a social system on the basis of its natives' self-image. However, how can we speak of 'real, active men' without speaking of what they 'say, imagine, conceive'? With this we return to the problem raised in the introductory chapter of this essay, and to Marx's assertion, there cited, of the determination of consciousness by social being. To quote the core of the 1859 Preface again:

> In the social production of their existence, men inevitably enter into definite relations, which are independent of their will, namely relations of production appropriate to a given stage in the development of their material forces of production. The totality of these relations of production constitutes the economic structure of society, the real foundation on which arises a legal and political superstructure and to which correspond definite forms of social consciousness. The mode of production of material life conditions the general process of social, political, and intellectual life. It is not the consciousness of men that determines their existence, but their social existence determines their consciousness.[10]

It will be useful to distinguish two, variously general, points at issue in this famous passage. The metaphor of foundation and superstructure is used, first of all, to express Marx's opposition to a particular approach to history. According to Hegel, the institutions of the political state provide the organising framework for social life as a whole. In the tradition of Enlightenment philosophy

---

9   Marx and Engels, *The German Ideology* (1976 [1845–46]), p. 36.
10   Marx, *Contribution* (1987 [1859]), p. 263.

of history, Hegel held that the development of the state was to be explained in terms of a logic of concepts – the dialectic – exhibited in the history of 'spirit', or the world of cultural categories. Marx's 'critical reexamination of the Hegelian philosophy of law' led him 'to the conclusion that neither legal relations nor political forms could be comprehended by themselves or on the basis of a so-called general development of the human mind, but that on the contrary they originate in the material conditions of life ...'. The 'anatomy' of these material conditions, Marx discovered, 'has to be sought in political economy'.[11]

However, Marx entitled the work in the Preface to which the above-quoted passages appear, a contribution not to political economy but to its critique. When he summarises the 'general conclusion' to which his economic studies led him, we find not the description of a new economic theory but the assertion that

> Just as one does not judge an individual by what he thinks about himself, so one cannot judge ... a period of transformation by its consciousness, but, on the contrary, this consciousness must be explained from the contradictions of material life, from the conflict between the social forces of production and the relations of production.[12]

This suggests an idea of greater generality than the rejection of legalist or mentalist explanations of social change. It suggests that in general the social theorist can and must distinguish between 'the economic conditions of production, which can be determined with the precision of natural science', and the 'ideological form' with which people represent these conditions and their changes to themselves.[13] And throughout the writings that constitute Marx's critique of political economy – the *Grundrisse*, the *Contribution, Capital, Theories of Surplus-Value*, and the mountain of notes and studies Marx left beside them – we discover that the categories of economics, whether in the form of economists' concepts or as they appear in 'the everyday consciousness of the agents of production themselves',[14] are treated as ideological forms, to be explained by reference to the social relations of production. Towards the end of the manuscript edited by Engels as Volume III of *Capital*, for instance, Marx refers to this ordinary consciousness as the 'religion of everyday life'.[15] While

---

11   Ibid., p. 262.
12   Ibid., p. 263.
13   Ibid.
14   Marx, *Capital*, Vol. III (1981 [1885]), p. 117.
15   Ibid., p. 969.

what Marx called 'vulgar economics' – the ancestor of today's 'neoclassical' economics – 'actually does nothing more than interpret, systematize, and turn into apologetics the notions of the agents trapped within bourgeois relations of production'.[16] These notions were to an extent critiqued by those whom Marx called the classical economists, above all by David Ricardo. But even these 'remain more or less in the grip of the world of illusion which their criticism had dissolved ...'.[17] This raises two interrelated questions: what is the relation of the critique of political economy to that science itself? and what are 'economic relations of production' if they are not to be identified with the referents of the categories of economic theory?

In his discussion of pre-capitalist society in the *Grundrisse* Marx observes that 'the statement that pre-bourgeois history, and each phase of it, has its own *economy* and an *economic basis* of its movement, is *au fond* merely the tautology that human life has from the beginning rested on production, and, *d'une manière ou d'une autre*, on *social* production, whose relations are precisely what we call economic relations'.[18] 'Production', here equated with economy, is what Marx in his methodological introduction to the *Grundrisse* called an *abstraction*, a concept which with respect to a group of phenomena 'brings out and fixes the common element and thus saves us repetition'. Insofar as it does this, such a concept is 'rational' (*verständig*). But it must not be forgotten that such abstractions are tools for organising the historical materials. Any particular case of production – of the 'metabolism between man and nature' – will be essentially characterised by a multitude of features. Some of these features will be common to all systems of production. But it is with production systems as with human languages:

> although the most highly developed languages have laws and categories in common with the most primitive ones, it is precisely what constitutes their development that distinguishes them from this general and common element. The determinations which apply to production in general must rather be set apart in order not to allow the unity which stems from the very fact that the subject, mankind, and the object, nature, are the same – to obscure the essential difference.[19]

---

16   Ibid., p. 956.
17   Ibid., p. 969.
18   Marx, 'Economic Manuscripts of 1857–58' (1986), p. 413.
19   Ibid., p. 23; see Evans-Pritchard on anthropological 'laws': 'Such generalizations, supported by a few selected illustrations, are either so general as to be devoid of significance or, where more precisely formulated, rest on too slender a base of evidence and fail to take

   Capitalism and feudalism both fall under the abstraction, 'system of production', but this is not to say that the properties of either system are to be elucid-
ated in terms of some unified set of characteristics. Each must be explained by
reference to the particular constellation of features ('determinations', in Marx's
vocabulary) characteristic of it as a unique historical object.

   Marx's 'abstractions' are not 'generalisations' suitable for the framing of universal laws linking social facts; their function is rather to provide a framework for comparison of social systems with respect to their differences. It may
be useful in particular to distinguish Marx's idea of 'abstraction' from Max
Weber's 'ideal type' concept. The latter is defined as an imaginary, pure case
of some pattern of motivation – the profit motive, for example, or the 'Protestant Ethic'. It is formed, in Weber's words, 'by the synthesis of a great many
... more or less present and occasionally absent *concrete individual* phenomena, which are arranged according to these one-sidedly emphasized viewpoints
into a unified *analytical* construct'.[20] Having no direct counterpart in reality,
they are 'primarily analytical instruments for the intellectual mastery of empirical data'.[21] Marx's abstractions, in contrast, are intended to refer to elements
of social reality, although their construction may involve processes of ideal-
isation. With respect to these concepts Marx is a realist as against Weber's
instrumentalism. 'Capitalism', for instance, refers for Marx to an actual social
structure. At any given moment (for example, at the time when Marx developed
the concept!) there may be no social system which as a whole instantiates
it. But this term refers to a developing system: the condition for its application is the existence of certain social relations which are developing over time.
The concept of 'capital' in fact assumes the existence of non-capitalist or pre-
capitalist social relations, since it is an historical concept; therefore it takes
account in advance of the fact that in any real case capitalist relations will be in
interaction with non-capitalist relations. All the category requires for its application is that the capitalist elements will dominate, either at the time or over a
time.[22]

---

        into account negative evidences ... The more the universality claimed, not only the more
        tenuous does the causal interpretation become but the more it loses also its sociological
        content' ('The Comparative Method in Social Anthropology' [1965], pp. 24–5).

20   Weber, *The Methodology of the Social Sciences* (1947), p. 90. In Weber's own opinion, 'all
        specifically Marxian "laws" and developmental constructs – insofar as they are theoret-
        ically sound – are ideal types', heuristically valuable but 'pernicious' 'as soon as they are
        thought of as empirically valid or real' (ibid., p. 103).

21   Ibid., p. 106. It should be noted that Weber takes this instrumentalist position with respect
        to all 'precise' concepts.

22   See Nowak, 'Weber's Ideal Types and Marx's Abstraction' (1978), pp. 81–91; for the rela-

What Marx has in mind when he speaks of the differentiation of cases falling under an abstraction by their differing complexes of determinations is
illustrated by his remarks on the category of 'population'. This term can be
taken to refer to a 'concrete', i.e. the empirical subject of social activity. But
theories of population, exactly because this abstraction is so general as to
refer to the members of every social system, shed no light on any particular
one.

> Population is an abstraction if, for instance, one disregards the classes
> of which it is composed. These classes in turn remain an empty phrase
> if one does not know the elements on which they are based, e.g. wage
> labour, capital, etc. These presuppose exchange, division of labour, prices,
> etc.[23]

The 'presupposition' and 'resting' here, reminiscent of the relation of foundation to superstructure, are certainly not meant ontologically; the relations
Marx has in mind are explanatory. As he later says, each involves the others:
'the simplest economic category, e.g. exchange-value, presupposes population,
population which produces under definite conditions; as well as a distinct type
of family, or community, or State, etc. Exchange-value cannot exist except as
an abstract, one-sided relation of an already existing concrete living whole'.[24]
Any human population, that is, exists ordered by specific social relations. To
understand a particular society, therefore, the system-specific definitions of
these relations must be studied. Furthermore, in such a study, the meanings
of the terms for these relations are given by their interrelation as elements of
the whole system.

The same holds true, in Marx's eyes, for 'production': the identity of all systems of production, or their elements – labour and means of production – is
proved by economists only 'by holding fast to the features common to all processes of production, while neglecting their specific differentiae. The identity
is demonstrated by abstracting from the distinctions'.[25] Marx makes the same
argument for terms denoting particular social institutions, when these institutions are common to several societies. Although money, for instance, plays a
crucial role in capitalism, it also existed 'in history before capital, banks, wage

---

tionship between concepts and history in Marx, see Morf, *Geschichte und Dialektik in der
politischen Ökonomie* (1970).

23   Marx, 'Economic Manuscripts of 1857–58' (1986), p. 37.
24   Ibid., p. 38.
25   Marx, 'Results of the Immediate Process of Production' (1976 [1863–66]), p. 982.

labour, etc. came into being'.[26] Marx analyses money as a commodity which plays the role of the form by which exchange-value is represented; this allows it to fulfil the functions of (to take two) the medium of exchange and the store of value. Both of these functions are to be discovered in pre-capitalist systems, but with a difference. In capitalism, money has the role of the 'universal equivalent': all goods with a central role in the reproduction of social life, including the basic good, labour power or the ability to produce, can be and are represented by sums of money. This is not true in pre-capitalist systems, where most goods are not produced for sale and where money does not mediate the social circulation of goods. Although both are cases of money, an explanation of money that fits one case may not be very helpful in answering questions about the other.

We thus encounter in Marx's critique of political economy a core element of Winch's critique of social science: the argument for the fallacious nature of equating features of different societies, made in terms of the systematic nature of social institutions. The unbridgeable gap between the two critiques would seem to be Marx's apparent disregard, in his desire for a scientific analysis of society, of the consciousness of the individuals constituting it.

On the other hand, we must note that Marx defines labour, the active factor in production, as 'purposive activity aimed at the production of use-values'.[27] Productive activity is of course itself regulated by consciousness; Marx stresses that his discussion of the labour process in *Capital* presupposes

> labour in a form in which it is an exclusively human characteristic ... [W]hat distinguishes the worst architect from the best of bees is that the architect builds the cell in his mind before he constructs it in wax. At the end of every labour process, a result emerges which had already been conceived by the worker at the beginning, hence already existed ideally. Man not only effects a change of form in the materials of nature: he also realizes his own purpose in the materials. And this is a purpose he is conscious of, it determines the mode of his activity with the rigidity of a law, and he must subordinate his will to it.[28]

Furthermore, Marx stresses that the aims of labour are socio-culturally and not naturally defined: 'Hunger is hunger; but hunger that is satisfied by cooked

---

26   Marx, 'Economic Manuscripts of 1857–58' (1986), p. 39.
27   Marx, *Capital*, Vol. I (1976 [1867]), p. 290; my emphasis.
28   Ibid., p. 284.

meat eaten with knife and fork differs from hunger that devours raw meat with the help of hands, nails and teeth'.[29] 'Food' itself, to develop this example, is too abstract to serve as a general name for the product of the relevant branches of production, while to treat it as meaning 'satisfier of biological need' is exactly to ignore the social determination of the humankind-nature metabolism. 'Food' denotes a use-value, it is true: that defined by the satisfaction of hunger. But in capitalism, according to Marx, use-values are the aim of production only insofar as they are commodities, objects saleable at a profit on the market. Food which cannot be sold will be destroyed. It is therefore produced not as food *per se* but as the commodity food, as a 'unity of use-value and exchange-value'. The general goal of production in capitalism is what Marx calls 'valorisation', the expansion in value of invested capital. Thus the labour process which abstractly appears as 'an appropriation of what exists in nature for the requirements of man, ... the universal condition for the metabolic interaction between man and nature'[30] in capitalism 'is only the means whereby the valorization process is implemented'.[31]

The distinction between abstraction and concrete example is twofold. On the one hand, since 'the requirements of man' are in part culturally as well as biologically defined, 'labour' in capitalism includes practices of no directly biological value – casting horoscopes, for example, or adjusting insurance claims. On the other hand, many activities such as brushing one's teeth, or even cooking one's own dinner, do not fall under the definition of labour relevant to the capitalist market system. Insofar as the two – humankind-nature metabolism and valorisation process – do coincide, capitalism and the physical requirements of human life are compatible; but if and when they don't, either the system or life must go, since 'production' at any time refers only to one particular instance of it or another.

Marx constructs a similar argument for the category of 'capital', persistently confused by economists with tools and materials in general. 'Whatever the social form of production', Marx writes in Volume II of *Capital*,

> workers and means of production always remain its factors ... For any production to take place, they must be connected. The particular form and mode in which this connection is effected is what distinguishes the various economic epochs of the social structure. In the present case, the

---

29   Marx, 'Economic Manuscripts of 1857–58' (1986), p. 29.
30   Marx, *Capital*, Vol. I (1976 [1867]), p. 290.
31   Marx, 'Results of the Immediate Process of Production' (1976 [1863–66]), p. 991.

separation of the free worker from his means of production is the given
starting point, and we have seen how and under what conditions the two
come to be united in the hands of the capitalist – i.e., as his capital in its
productive mode of existence.[32]

Thus, to take the classical economists' concept of 'fixed capital', 'means of
labour are fixed capital only where the production process is in fact a capitalist process and the means of production are thus actually capital, i.e. possess
the economic determination, the social character of capital; secondly, they
are fixed capital only if they transfer their value to the product in a particular way'.[33] The abstract definition of 'means of production' as 'objects used
in the process of producing other objects' does not hold true for 'fixed capital', in the strict sense that being such an object is necessary but not suffi-
cient for such an object's being capital. A tool can only function as capital
if its use makes possible the exploitation of labour power and the eventual
generation of a sum of money exceeding that invested in the production process. Thus the abstraction, 'means of production', applies to capital only in so
far as 'production' is here taken in its special form as the production of *capital*.

But such definitions, which Marx is opposing to those of the economists,
fundamentally involve reference to cultural categories: as Marx puts it in the
passage just cited, 'capital' (for instance) is a social character or determination. 'Capital' is defined as self-expanding 'value', and 'value' in terms of the
socially necessary labour time needed to produce commodities. This labour
time appears to refer to a natural phenomenon, the hours of actual work put
into producing a given object, and is generally taken to do so by economists
commenting on Marx, but it does not. Marx makes quite a point of distinguishing between labour as use-value-creating, as natural phenomenon, and labour
as value-creating, as socially-defined 'abstract labour'. The latter is defined
only in the context of the sale of goods on the market.[34] 'Commodity', once
it becomes the general form of goods, turns out not to be disentangleable from
the category of 'capital', while 'exchange' itself requires – as noted in the first
chapter of this book – reference to a social mode of behaviour in which people

---

32   Marx, *Capital*, Vol. II (1978 [1884]), p. 120.

33   Ibid., p. 303. See Marx, 'Results of the Immediate Process of Production' (1976 [1863–66]),
       p. 909.

34   See Mattick, 'Some Aspects of the Value-Price Problem' (1981), pp. 725–81 [an improved
       (and significantly corrected) version of this article can be found in Mattick, *Theory and
       Critique* (2018), Ch. 6.]

recognise a category of 'ownership' and 'behave in such a way that each does not appropriate the commodity of the other, and alienate his own, except through an act to which both parties consent'.[35]

The concept of 'capital' is thus *ideological* in the same way as the concept of 'nobility' discussed in the previous chapter: by conflating the specific, historical form of the means of labour with 'means of labour' as a transcultural abstraction, the particular institutions of capitalist society are presented as universal, as present, even if in different disguises, in all societies. We will return to this example below. Here I want to point out that Winch is wrong to assimilate Marx's concept of ideology to Pareto's 'derivations'. This is because Marx's abstractions are not 'residues', real contents hidden by ideological appearances. For Pareto, the various cases that represent the same residue are equivalent to each other, except for their superficial appearances. This is exactly what Marx wishes to deny in the case of abstractions. The differences that are extinguished in the formation of abstractions are what we must appeal to in any attempt to explain the features of actual societies.

> The materials and means of labour ... play their part in every labour process in every age and in all circumstances. If, therefore, I label them 'capital' ... then I have *proved* that the existence of capital is an eternal law of nature of human production, and that the Kirghiz who cuts down rushes with a knife he has stolen from a Russian so as to weave them together to make a canoe is just as true a capitalist as Herr von Rothschild. I could prove with equal facility that the Greeks and Romans celebrated communion because they drank wine and ate bread, and that the Turks sprinkle themselves daily with holy water like Catholics because they wash themselves daily.[36]

This argument holds even for a case like that of 'labour', which unlike 'capital' is an abstraction like 'production', denoting human effort in metabolising nature, and so is valid for all social systems. On the other hand, it has a specific use for the analysis of capitalism which it does not have for any other system. It is only in this society that the idea of a 'labour force', the individual members of which 'easily pass from one kind of labour to another',[37] has operational meaning, since in earlier systems the division of labour was more strictly regulated

---

35   Marx, *Capital*, Vol. I (1976 [1867]), p. 178.
36   Marx, 'Results of the Immediate Process of Production' (1976 [1863–66]), p. 991.
37   Marx, 'Economic Manuscripts of 1857–58' (1986), p. 41.

by custom. In addition, all forms of labour are treated as equivalent in capitalism, in the sense that their embodiments in products are equally transformable into money. In contrast, when applied to pre-capitalist systems, 'labour' changes its significance. It cannot function as a category for the specific form that production may take in any given society either. We must distinguish, that is, between 'labour' as an abstraction denoting productive activity in any society and 'labour' as a category for such activity in a particular society, capitalism. A better-known example, which has proved equally as troublesome for commentators on Marx, is the concept of 'class'. In describing 'all previous history' as 'the history of class struggles', Marx uses 'class' to designate an abstraction: social grouping defined by its control over the means of social production. When describing the evolution of capitalism out of pre-capitalist society, in the *German Ideology*, however, he draws attention to the difference between class and estate as forms of social organisation, pointing out the historical specificity of the former to capitalism.

Marx seems indeed to have held a conception closer to Winch's idea of a social science than the latter suspects. But here we must remember the double use Marx makes of his metaphor of foundation and superstructure: to make not only the specific contrast between processes of (consciously regulated) interaction with nature and other activities, but also the more general contrast between social institutions and people's understanding of them. To say that the processes of social life are structured by culturally-determined categories is not – as we saw earlier – to say that people's understanding of these processes cannot go seriously awry. Although it is by their belief in witchcraft that the Azande create an object of study for the outsider anthropologist, this does not mean that the phenomena categorised as witchcraft and magic are correctly described or understood by the insiders. As Zande witchcraft for Evans-Pritchard, so does capitalist economics serve as the subject matter of Marx's study, which is thus an example of ideological explanation in the second of Mackie's senses and not the first (see p. 53 above).

In effect, Marx's work threatens a classic distinction between anthropology and the other analytic social disciplines. S.F. Nadel, for example, finds no difference in principle between anthropology and sociology, but distinctions in method and concept-formation arising from the character of the objects of anthropological investigation – the 'strangeness of primitive cultures, their separation from our civilization':

> Sociology as well as history presuppose a certain familiarity of the observer with the nature of the data he studies; both disciplines imply that certain commonly used categories are immediately applicable, and that

the probabilities and consequences of actions are either readily understood or deducible with self-evidence from the general background of
experience. New items ... or new problems ... fall into place against this
background of an experience which is common to observer and observed.
This is not true of primitive society, which is unfamiliar and often strange
... Social categories must be evolved *ad hoc*; they cannot simply be taken
for granted.[38]

As Nadel observes, such an approach need not be limited to strange cultures;
one may go further to ask how it could be, if the goal is something we would call
a 'scientific' understanding of social life. The contrast with the anthropological
point of view can be brought out if we look back to the Martian whom Fritz
Machlup had investigating the stock market (see Chapter 2 above). In Nadel's
terms, Machlup's Martian is not doing anthropology but economics. He is trying to explain the functioning of the stock market within the context of the
economic and social system as a whole, and indeed the concept-system and
culture of capitalism itself. This is implicit in Machlup's statement of the Martian's goal as 'to know the economic functions of the stock market, particularly
its role in the utilization of investible funds and in the formation of capital'.[39]
Such a goal is imaginable only if there is capital investment, and all that goes
with it, on Mars!

Machlup's questions are anthropological ones only as the demand for ideological explanation of Mackie's first sort is, as explorations of the inner articulation of the culture studied. Machlup imagines these to be anthropological questions of a deeper sort perhaps because he is himself an economist, for whom
economic theory consists of a system of '"universal laws" or fundamental hypotheses' involving 'constructs of idealized human action based on [assumed]
objectives to maximize profits and satisfactions'.[40] In line with this Weberian
definition of his field, Machlup sees the theorising necessary for understanding
the market as consisting 'mainly in constructing ideal types of motivated conduct of idealized decision-makers, and examining them in abstract models of
interactions'. So, for example, business competition is to be explained in terms
of a 'profit motive'. Of course, these motivations can only exist in a social system
in which the requisite categories of action – market exchange, capital, profit –
exist; Machlup focuses on the goal-defined ideal types of action because he

---

38   Nadel, *The Foundations of Social Anthropology* (1953), p. 5.
39   Machlup, 'If Matter Could Talk' (1969), p. 297.
40   Ibid., p. 300.

takes the social categories for granted. His anthropologist from Mars is really an economist from Earth in a space suit.[41]

Were he actually to make the effort, in Nadel's words, 'to treat a familiar culture as though it were a strange one', the anthropologist would go beyond the analysis of a given institution or phenomenon in terms of capitalism's own categories ('profit', 'market', 'capital', etc.) to an explication of these categories themselves. He would in this process be forced to evolve new categories 'ad hoc', in the attempt to develop a theory about those structured activities that in contemporary society we insiders think of as 'the economy', a theory which might ultimately provide an explanation and analysis of this classification itself. Thus an anthropologist would be led to ask why instruments of production should be construed under the rubric of 'capital', i.e. as equivalent to certain sums of money, just as Evans-Pritchard wondered about the equivalence of birds and people among the Nuer and Willett about the Yoruba's identification of cowrieshell boxes with their heads or souls.

For the economic theorist the most abstract categories are those of his own culture; thus economic anthropologists have struggled – without notable success – to impose the concepts of neoclassical economics on their field studies of tribal societies. In reality, in other cultures (for instance that of medieval Europe), the category of 'capital' – and therefore a concept like 'profit-maximisation' – plays no real part at all, any more than the concept of inherited witchcraft plays a role in the culture of capitalism. The anthropologist's task (as stated concisely by Julian Pitt-Rivers), if he is not 'to accept the "natives"' view

---

41   The distinction is apparently clear at least to Martian academics, as witness an editorial, entitled 'The Martian's View', from the *International Herald Tribune* of 5 February 1982:

The man from Mars dropped into earth orbit, just in time. He had heard that the first Reagan budget was about to appear. What, he asked, was the focus of public attention? ... The Martian wanted to know whether the [national deficit] number turned out to be accurate. We laughed, and explained that everyone knew it would be pretty fake from the beginning ...

'If everybody knew the number was bent', the man from Mares asked, 'why did the president go to such lengths to produce it?' He kept asking questions like that. You could tell he was from Mars.

'It makes people feel better', we patiently explained. 'That's why Mr. Reagan is going to struggle so hard ... to keep his deficit figures under $100 billion. It's a matter of paying respect to the proprieties, like the medicine man doing the rain dance ...'

If the budget office puts out a fairly reliable set of numbers, the Martian asked, why does the president put out different ones?

'That's politics', we said, 'which would hardly interest a serious economist like you'.

'I'm not an economist', the Martian indignantly exclaimed. 'I'm an anthropologist. I'm writing a book about the tribal habits of the smaller planets'.

of their own world as if it bore the credentials of science', is 'to re-order that which they order and discover behind their ordering a more abstract order that explains it'.[42] It is the inability to do this – to adopt the outsider's point of view in the study of their own culture – which typically mars the curious attempts of anthropologists to apply their techniques to capitalist society. Almost always such studies examine an area, a neighbourhood, or an occupational grouping.[43] But if the classic topics of anthropology are such as the Nuer, the Mbuti, the Nayars, or even 'tribal society' globally, then the equivalent frame of reference for studying modern society should also be the whole culture – the capitalist world system. For this reason one may speak of Marx's work as anthropological in approach, rather than as sociological or economic (although it would be misleading to speak of Marx as an anthropologist, since for him scientific understanding of capitalism was to be only a means to its revolutionary abolition).

As a matter of fact, Marx contrasts his theoretical procedure with that of the economists in the same terms as those we have used to contrast anthropological investigation with the researches of Machlup's pseudo-Martian. The starting point of Marx's research procedure in *Capital* is an attempt to see the categories of his own culture as foreign and strange. 'A commodity appears at first sight an extremely obvious, trivial thing. But its analysis brings out that it is a very strange thing, abounding in metaphysical novelties and theological niceties'.[44] The language of 'metaphysics' and 'theology', and above all the dominant metaphor of 'fetishism' is not accidental or purely literary: Marx means to suggest that the analysis of economic ideology is a case of the same proced-

---

42   Pitt-Rivers, 'On the Word "Caste"' (1971), pp. 231–2.

43   Eric Wolf's position is typical: 'The anthropologist's study of complex societies receives its major justification from the fact that such societies are not as well organized and tightly knit as their spokesmen would on occasion like to make people believe ... [The] formal framework of economic and political power exists alongside or intermingled with various other kinds of informal structure which are interstitial, supplementary, parallel to it ... The anthropologist has a professional license to study such ... structures in complex society and to expose their relationship to the major strategic, overarching institutions' ('Kinship, Friendship, and Patron-Client Relations in Complex Societies' [1966], p. 1). 'Professional licenses' are clearly parcelled out by reference to academic fields, rather than by the nature of social subject matter; anthropology's claim to some bits of modern social reality are necessitated by the disappearance of the 'simpler' societies that were its original preserve. On a more conceptual level, what we see here is the projection onto modern society of categories evolved (within that society) in the study of 'simpler' ones, just as economic anthropologists attempted to examine 'primitive' economies with the categories of marginal utility theory.

44   Marx, *Capital*, Vol. I (1976 [1867]), p. 163.

ure as is represented by the analysis of religious or philosophical ideology. The 'strangeness' of the commodity is not noticed by the economists who study it. This is because, in Marx's words,

> Reflection on the forms of human life, hence also scientific analysis of those forms, takes a course directly opposite to their real development. Reflection begins *post festum*, and therefore with the results of the process of development already to hand. The forms which stamp products as commodities already possess the fixed quality of natural forms of social life before man seeks to give an account, not of their historical character, for in his eyes they are immutable, but of their content and meaning ...
>
> The categories of bourgeois economics consist precisely of forms of this kind. They are forms of thought which are socially valid and therefore objective [are 'social facts', to use Durkheim's terms] for the relations of production belonging to the historically determined mode of production, i.e., to commodity production. The whole mystery of commodities ... vanishes therefore as soon as we come to other forms of production[45] – i.e., as soon as we put them into a comparative perspective.

This is of course one of the clichés of anthropology, that cross-cultural comparison demonstrates the particularity of what appears to the cultural insider to be universal. But, as Winch objects, how can we compare that which is culturally specific? To quote Pitt-Rivers again,

> A comparison of social structures must be concerned with the elements of social structure: hierarchy, authority, power, sanctity, the division of labor, the definition and solidarity of groups, the transmission of property, the rules of descent, etc., not with approximations to a stereotype, that is to say, a particular conjunction of features which may frequently be associated in fact but have no demonstrated necessary relationship to one another.[46]

Here lies the role of the Marxian abstraction: the subordination of 'capital' (e.g.) to the complex of abstractions 'tool use' and 'exploitation' (to take the most important elements of Marx's explication of this institution) exposes the par-

---

45   Ibid., pp. 168–9.
46   Pitt-Rivers, 'On the Word "Caste"' (1971), p. 250.

ticularities of the capitalist mode of production. One can even say that it is, as Winch recognises in his attack on transcultural analysis, the availability of comparison that defines the outsider's point of view (as opposed to being an actual traveller from a different society).

How can this be? How, that is, could anyone and how can Marx in particular be said to have had an outsider's point of view on what was after all his own social system? The answer for the general case lies in the fact that modern society certainly, and most societies to some extent, far from being the monoliths suggested by our use of the term so far (which follows Winch's) are complexes of different interests, or (to use the jargon expression) subcultures. Marx wrote as a member of the socialist movement of the second half of the nineteenth century, which by that time had already developed categories defining a new form of social life – based on the abolition of wage labour and the substitution for it of the 'free association of producers' – incompatible with the institutions of capitalism. The novelty of capitalism in world history is that it is a class society in which one class represents an entire potential social system; it is this which was expressed by the socialist movement. In this context Marx was an outsider within the system itself, in that he could look at it from the perspective of a society yet to be created, *but already represented by new categories of social action.*

Similarly, Marx claimed that the categories of capitalist culture come to be used to understand other (earlier) forms of society only when capitalism itself is seen as an historically specific form of society, rather than as either the equivalent of other forms or the endpoint of historical development.[47] The scientific strength of classical political economy derived also from its peculiar historical position at a period when a new form of society, capitalism, was still in conflict with remains of the pre-capitalist system. (This is visible, for example, in Smith's polemic against mercantilist policy and for laissez-faire.) Even though the classical thinkers, true representatives of the Enlightenment, made their arguments in terms of the 'naturalness' of capitalist institutions, they still in practice contrasted those institutions with the leftovers of the earlier order. This implicit conflict of points of view, then, brought into focus the specific structure of the new system; so that, 'although encompassed by [the] bourgeois horizon, Ricardo analyses bourgeois economy, whose deeper

---

47 '[I]t was not until the self-criticism of bourgeois society had begun that bourgeois [political] economy came to understand the feudal, ancient and oriental economies' (Marx, 'Economic Manuscripts of 1857–58' (1986), p. 43). 'The owl of Minerva spreads its wings only with the falling of the dusk'.

layers differ essentially from its surface appearance, with such theoretical acumen that Lord Brougham could say of him: "Mr Ricardo seemed as if he had dropped from another planet."[48]

Whether or not this explanation for the anthropological character of Marx's study of economics is correct, it is certainly true that this is one way of explaining why he called his writings a *critique* of political economy. His position was, in Marx's own opinion, ably summed up by I.I. Kaufmann, the first Russian reviewer of *Capital*, who wrote that the position that 'the general laws of economic life are one and the same, no matter whether they are applied to the present or the past,'

> is exactly what Marx denies. According to him, such abstract laws do not exist … On the contrary, in his opinion, every historical period possesses its own laws … The old economists misunderstood the nature of economic laws when they likened them to the laws of physics and chemistry. A more thorough analysis of the phenomena shows that social organisms differ among themselves as fundamentally as plants or animals. Indeed, one and the same phenomenon falls under quite different laws in consequence of the different general structure of these organisms, the variations of their individual organs, and the different conditions in which these organs function.[49]

Marx is therefore with Winch in opposition to the idea of social science as aiming at the discovery of general laws of social life. What general 'laws' are frameable in this realm are reducible 'to a few very simple definitions, which are expanded into trivial tautologies'.[50] Similarly, the relation between social consciousness and relations of production, between superstructure and foundation, cannot be stated except by the most abstract of formulas, so that 'in order to examine the connection between intellectual production and material production it is above all necessary to grasp the latter itself not as a general category but in *definite historical* form'.[51] Agreement with Winch ends, however, at the conception of this 'definite historical form'. For Winch cultural forms are wholly defined by the meanings they represent for natives, so that understanding society is 'grasping the *point* or *meaning* of what is being done or said'.[52] For

---

48   Marx, *Contribution* (1987 [1859]), p. 300.
49   Marx, *Capital*, Vol. I (1976 [1867]), p. 101.
50   Marx, 'Economic Manuscripts of 1857–58' (1986), p. 24.
51   Marx, 'Economic Manuscript of 1861–63' (1989), p. 182.
52   Winch, *Idea* (1963), p. 115.

Marx, in contrast, 'grasping the point or meaning' of cultural items for the natives is only the prelude to the knowledge he is seeking: it is only to identify that which is to be explained, that whose meaning for the scientific investigator is to be established.

For Max Weber, as we saw, complete explanation requires the conjoining of a motivational interpretation with observed statistical regularity of behaviour. As we noted in Chapter 2, Winch quite correctly points out that if we construe meaning as something 'subjective' it is hard to see how correlations with statistical regularities can justify interpretations. Therefore, of necessity, in attempting to describe a social situation in natural law-like terms,

> Weber ceases to use the notions that would be appropriate to an interpretive understanding of this situation. Instead of speaking of the workers in his factory of being paid and spending money, he speaks of their being handed pieces of metal, handing those pieces of metal to other people and receiving other objects from them; he does not speak of policemen protecting the workers' property, but of 'people with helmets' coming and giving back the workers the pieces of metal which other people have taken from them; and so on. In short, he adopts the external point of view and forgets to take account of the 'subjectively intended sense' of the behavior he is talking about: and this, I want to say, is a natural result of his attempt to divorce the social relations linking those workers from the ideas which their actions embody: ideas such as those of 'money', 'property', 'police', 'buying and selling', and so on. Their relations to each other exist only through these ideas and similarly these ideas exist only in their relations to each other.[53]

Marx would be in full agreement with the bulk of this passage. Gold is certainly not (in his eyes) to be identified with money, nor are all authorities with helmets policemen. On the other hand, he strenuously denies that social relations exist only through ideas, since the social relations identified by their ideal representations among the natives, being observable, are open to different descriptions by outside observers. The error shared by Winch and Weber lies in the assumption that the alternatives of 'objective' description and 'subjective' interpretation exhaust the possibilities of social understanding. As we saw, this is not true.

---

53   Ibid., pp. 117–18.

In the previous chapter I sketched a distinction between 'common-sense' and 'theoretical' thinking about social life. The first is represented by Azande thinking about witchcraft in response to Evans-Pritchard's questions, and by the stock market workers' conceptions of their roles in the economy. The second is represented by Evans-Pritchard's systematisation of Azande doctrine and by economic theory. Here Winch draws the line of acceptability, for the third sort of thinking about social life, which Marx called specifically scientific theorising, represents a theoretical construction on the native categories which the natives themselves would not recognise as valid.

Such an approach is justified by Marx as affording answers to questions that insiders will not ask. Vulgar economy, as Marx sees it, merely 'interprets, systematizes, and defends' the going conceptual scheme of (social) things. 'But all science would be superfluous if the form of appearance of things directly coincided with their essence'.[54] Here Marx adapts the distinction of essence and appearance to the social world: the phenomena to be explained by social theory include the appearances of the system in the eyes of its inhabitants. The distinction between the two, as Marx conceived it, can be seen in his discussion of the dual nature of Adam Smith's work in *The Wealth of Nations*:

> On the one hand he attempted to penetrate the inner physiology of bourgeois society, but on the other, he partly tried to describe its externally apparent forms of life for the first time, to show its relations as they appear outwardly, and partly he even had to find a nomenclature and corresponding mental concepts for these phenomena ... for the first time ... The one task interests him as much as the other, and since both proceed independently of one another, this results in completely contradictory ways of presentation: the one expresses the intrinsic connections more or less correctly, the other ... expresses the *apparent* connections [between economic phenomena]

– apparent, that is, 'to the unscientific observer just as to him who is actually involved ... in the process of bourgeois production'.[55] Smith was engaged in systematic description of different aspects of the capitalist social system, in terms of the everyday categories of 'value', 'price', 'rent', etc. At the same time, when he used the concept of 'labour' to explain the phenomenon of value,

---

54   Marx, *Capital*, Vol. III (1981), p. 956.
55   Marx, 'Economic Manuscript of 1861–63' (1989), pp. 391, 390.

he made use of a category which, as used in his theory, was not part of the vocabulary of everyday life.

But, in Marx's opinion, despite the efforts of the classical economists, even they 'remained more or less trapped in the world of illusion … and nothing else is possible from the bourgeois standpoint; they all fell therefore more or less into inconsistencies, half-truths, and unresolved contradictions'.[56] Just because in capitalism the values of commodities are determined by 'the market', and confront the owners of the commodities as facts limiting their choices of action, the determination of exchange-values appears as the outcome of natural laws of economics. The fundamental categories of these laws – 'commodity', 'value', and (for classical economics) 'labour' – appear to the economists as basic properties of the social world.

> Political economy has indeed analyzed value and its magnitude, however incompletely, and has uncovered the content concealed within these forms. But it has never once asked the question why this content has assumed that particular form, that is to say, why labor is expressed in value …[57]

Why, that is, is labour represented in this system by price, the exchange-value of commodities relative to the money commodity? Marx's answer is to explain the category of value, the property of goods measuring their mutual exchangeability, in terms of specific features of capitalist society. Social labour appears as value, i.e. is organised through the market, (a) when labour has become radically social, in the sense that production is carried on for society at large rather than for the producer or immediate social groups; (b) while control over property remains individual; (c) and labour itself is treated as a good to be exchanged in the market against other things. This condition requires the separation of the labouring population from the means of production, without which they cannot produce for themselves. We may distinguish two sides to this explanation. First, the category, 'value', is explained as the representation in this culture of a certain organisation of social production. Second, this organisation is explained as the outcome of a particular historical process, and not as a consequence of some 'inner nature of production itself'. While it is an explanation, such an account is not a substitute for or a translation of the economic categories. Just as life in Zande society required Evans-Pritchard to talk in terms

---

56   Marx, *Capital*, Vol. III (1981), p. 969.
57   Marx, *Capital*, Vol. I (1976), pp. 173–4.

of magic and witches, despite his disbelief in them, life in capitalism requires that one does sell one's labour power, or buy someone else's, and so think of labour as a commodity. And

> If I state that coats or boots stand in a relation to [gold] because the latter is the universal incarnation of abstract human labour, the absurdity of the statement is self-evident. Nevertheless, when the producers of coats and boots bring these commodities into a relation with ... gold or silver ... as the universal equivalent, the relation between their own private labour and the collective labour of society appears to them in exactly this absurd form.[58]

'Value' does not *mean* 'socially necessary abstract labour time', although that, according to Marx's theory, is what it is. The latter concept cannot replace the former in economic life; it would have no significance there. It is because, under the existing conditions of private appropriation of the social product, the abstract or social character of labour can have no direct form of representation (as it would, for example, in a system in which decisions about social production were collectively made), that social labour can only be represented, as 'value', in the form of money.

The Marxian analysis of money, as 'the commodity whose natural form is also the directly social form of realization of human labor in the abstract',[59] is not open to the objection made by Winch to Weber's attempt at an 'objective' description. Money is defined in terms of the cultural institutions referred to by the categories of 'commodity', 'price', and so on. At the same time, Marx's account also escapes the alternatives posed by Winch, of cultural relativism or a priori truths of social life, by explaining both what the essence of this phenomenon is and why it appears as it does to the natives of capitalism.

To illustrate this, I will return to the analysis of the category of 'capital', money used to make money. According to Marx, 'capital is not a *thing*, any more than money is a *thing*. In capital, as in money, ... certain social relations appear

---

58   Marx, *Capital*, Vol. I (1976), p. 169. Compare Evans-Pritchard's remarks in *Witchcraft*: 'It is difficult for us to understand how poison, rubbing-board, termites, and three sticks can be merely things and insects and yet hear what is said to them and foresee the future and reveal the present and the past, but when used in ritual situations they cease to be mere things and mere insects and become mythical agents' (p. 151). This is of course a prime example of the phenomenon of fetishism, of which Marx believed the power of gold to incarnate human labour was another.

59   Marx, *Capital*, Vol. I (1976), p. 241.

as the natural properties of things in society'.[60] 'Capital' defines things as they fit into a pattern of social action.

> Under certain circumstances a chair with four legs and a velvet covering may be used as a *throne*. But this same chair, a thing for sitting on, does not become a throne by virtue of its use-value. The most essential factor in the labour process is the worker himself, and in antiquity this worker was a *slave*. But this does not imply that the worker is a slave by nature ..., any more than spindles and cotton are *capital* by nature just because they are consumed nowadays by the *wage-labourer* in the labour process. The folly of identifying a specific *social relationship of production* with the thing-like qualities of certain articles is what strikes us most forcefully whenever we open any textbook on economics and see on the first page how the elements of the process of production, reduced to their basic form, turn out to be land, *capital*, and labour ... By confusing the appropriation of the labour process by capital with the labour process itself, the economists transform the *material elements* of the labour process into capital simply because capital itself changes into the material elements of the labour process *among other things*.[61]

Marx does not stop, as Winch would wish, at noting this social definition. First, Marx judges this description of the economists as ideological, in Mackie's sense. It contains the error of treating a social character as a natural property. Marx assimilates this error to that of attributing an independent power to man-made idols by speaking of the 'fetishism of commodities', i.e. the treatment of the results of a stage of social development as ineluctable laws to which humans can only submit.

---

60   Marx, 'Results', p. 1005; see p. 1003: 'Man can only live by producing his own means of sub-
      sistence, and he can produce these only if he is in possession of the means of production,
      the material conditions of labour. It is obvious from the very outset that the worker who
      is denuded of the means of production is thereby deprived of the means of subsistence,
      just as, conversely, a man deprived of the means of subsistence is in no position to create
      the means of production. Thus even in the first process, what stamps money or commod-
      ities as capital from the outset, even before they have really been transformed into capital,
      is neither their money nature nor their commodity nature, nor the material use-value of
      these commodities as means of production or subsistence, but the circumstance that this
      money and this commodity, these means of production and these means of subsistence
      confront labour-power, stripped of all material wealth, as autonomous powers, personi-
      fied in their owners'.
61   Marx, 'Results', pp. 997–8.

Second, he explains the ideology, by showing what aspects of capitalist social experience (as described in his own theoretical vocabulary) call for this ideological interpretation. In capitalism,

> Raw materials and the object of labour in general exist only to *absorb* the work of others, and the instrument of labour serves only as a conductor, an agency, for this *process of absorption* ... Since work creates value only in a definite useful form, and since every particular useful form of work requires materials and instruments with specific use-values, ... labour can only be drained off if capital assumes the shape of the means of production required for the particular labour process in question, and only in this shape can it annex living labour. This is the reason, then, why the capitalist, the worker, and the political economist, who is only capable of conceiving the labour process as a process owned by capital, all think of the *physical* elements of the labour process as *capital* just because of their physical characteristics. That is why they are incapable of detaching their physical existence as mere elements in the labour process from the *social* characteristics amalgamated with it, which is what really makes them *capital.* They are unable to do this because in reality the labour process that employs the physical qualities of the means of production ... is identical with the labour process that converts these self-same means of production into capital.[62]

The sorry state of economic theory mentioned in Chapter 1 of this essay, in combination with the most severe economic difficulties since the thirties, has led to a crisis of economics serious enough for the Nobel and Rockefeller foundations to invest millions in a study of the conceptual bases of economic policy, described by Dr Victor Urquidi as 'theoretical concepts that have nothing to do with the present or looking to the future'.[63] We may note also a rash of books and articles on the insufficiencies of economics. However, the economists are neither resigning their academic and governmental posts nor being asked to do so. As one journalist commented, 'The economists have a great deal invested in their present "scientific" approach, both intellectually and practically – in professorships, learned journals, possession of graduate students, positions in government and industry, honors, up to and including the Nobel Memorial Prize

---

62   Ibid., pp. 1007–8.
63   'Economic Theories Face Test', *New York Times*, 21 June 1983.

in Economic Science'.[64] And their employers must find the thought of being without scientific advice, counsel, and policies to sponsor equally disagreeable.

True though this explanation is, there is more to the matter than this; we must explain the material success of economic theory, given its scientific failure. A factor that springs to mind is the apparent success obtained by Keynesian economics during the 1950s and '60s in managing the business cycle. But not only did this success turn to stagflationary ashes in the mouths of its pundits; more significant is the fact that throughout the heyday of deficit financing neo-classical economics continued to be taught and accepted as an integral part of economic science, despite the proof of its own inadequacy presented by the Great Depression, which had launched an earlier 'crisis of economic theory'. This situation, indeed, is reminiscent of nothing so much as the disinterest of the Azande in exploring and resolving the contradictions in their account of witchcraft. 'Azande do not perceive the contradiction as we perceive it, because they have no theoretical interest in the subject, and those situations in which they express their beliefs in witchcraft do not force the problem upon them', wrote Evans-Pritchard. Here, however, we see the over-optimism in Barden's idea that theoretical thinking has a natural tendency to become critical: while for the economists their subject is a theoretical one, they nonetheless have little interest in its methodological presuppositions, given that the academic or policy advisory situations in which they express their beliefs in economics do not force such problems upon them.

Marx's theory of capitalist society, as we have seen, goes far towards explaining this state of affairs. If economic categories play as central a role in the definition of capitalist social relationships as witchcraft categories do in the life of the Azande, it is not hard to understand the resistance of this network of concepts to the effects of theoretical incoherence and empirical disconfirmation. What drew Marx's attention to political economy as an object for critical analysis was the combination of its theoretical inadequacy and its dominion over professional intellectual and ordinary thinking alike. He was thus faced with

---

64 Silk, 'Economic Scene', *New York Times*, 8 July 1983. Compare Lester Thurow's article, 'It's All Too Easy to Be a Critic', *New York Times*, 19 June 1983. Among the most important criticisms of economic theory from within the profession, see Shoeffler, *The Failures of Economics* (1955); Morgenstern, *On the Accuracy of Economic Observations* (1963) and 'Thirteen Critical Points in Contemporary Economic Theory' (1972), pp. 1163–89; Ward, *What is Wrong With Economics?* (1972); Routh, *The Origin of Economic Ideas* (1975); Canterbery and Burkhardt, 'Economics: The Embarrassed Science' (1979); and Eichner (ed.), *Why Economics is Not Yet a Science* (1983).

the same problem as Evans-Pritchard: the explanation of a conceptual scheme that is functionally indispensable to the life of a culture despite its inconsistencies and absurdities.

The adequacy of Marx's theory of capitalist society and its forms of consciousness cannot be a topic for the present essay. But it should be noted that since this theory aims not at interpretation of capitalist culture, in the sense of a translation of capitalist into non-system-specific categories, but at an explanation of the continuing existence of this social system, its adequacy is not to be judged by its conformity to 'the ordinary consciousness of the agents of production themselves'. Rather, the theory's strength is to be measured, first, by its ability to explain that 'ordinary consciousness' and, second, by Marx's success in explaining and predicting the development of a social system structured by the categories of capital and wage-labour. As I have indicated, I find *Capital*, judged in these terms, to contain the most important well-confirmed theory within social science. Whatever the difficulties from which Marx's theory may prove to suffer, however, they cannot be shown to derive either from neglect of the intentional character of social categories or from a failure to observe the canons of scientific objectivity.

CHAPTER 6

Science and Society

I have argued in this essay that the 'meaningful' character of 'social facts' does not put them beyond the reach of scientific explanation. On the contrary, it is the very possibility of scientific understanding of society that makes it important to reject many of the claims currently made to this title. The many failures of official social science are due not to an inherent resistance to scientific inquiry on the part of social phenomena, but to the would-be scientists' failure to include their own culture in the domain of investigation.

Of course, it does not seem to them that they have done so. Economics, for instance, has explicitly sought to discover highly abstract laws applying to all social systems, just as sociologists have striven to unveil general principles of human behaviour applicable to the visible spectrum of such behaviour. Yet the former discipline, to take what is generally regarded as the 'queen of the social sciences', remains 'in the grip of the world of illusion' constituted by the categories of everyday life. This judgement of Marx's has the virtue of explaining the explanatory sterility of economics and its failure as an empirical predictor and basis for policy choice. Fundamentally, as we have seen, the root of the problem is to be found in the goal of discovering general laws of social life. These general laws turn out upon examination to be merely the projection upon the totality of social experience of the categories of modern society. It is the inability to see what Karl Korsch called the 'historical specificity' of these categories that has blocked even recognition of Marx's discovery of the principles regulating capitalism as a social system.[1]

It has been held – despite his approval of his Russian reviewer's criticism of the idea – that Marx himself maintained a general theory of society ('historical materialism'), of which his account of capitalism is then the application to a particular case. In the 1859 Preface to *Zur Kritik* Marx affirms in the tradition of European 'universal history' that 'Asiatic, ancient, feudal, and modern bourgeois modes of production may be designated as epochs marking progress in the economic development of society', with bourgeois society as 'the last antagonistic form of the social process of production'.[2] This way of speaking derives from the Enlightenment's periodisation of history, as preserved and

---

1 [See Korsch, *Karl Marx* (1963 [1938]), Chs. 2 and 3.]
2 Marx, *A Contribution to the Critique of Political Economy* (1987 [1859]), pp. 263–4.

© KONINKLIJKE BRILL NV, LEIDEN, 2020 | DOI:10.1163/9789004414822_007

transformed – *aufgehoben* – by Hegel. While for the Enlightenment human nature was constant throughout history, with the advance from one historical stage to another marked by the growth of knowledge of that nature and the laws governing it, Hegel developed the Romantic idea of a progression of the human spirit itself, through a series of civilisations, each with its own unifying principle. This view is related more closely than its predecessor to the modern anthropological concept of culture; it is even an ancestor of Winch's view of cultures as monad-like systems of meaning. Commensurability of cultures is fundamental to the Hegelian conception, however, since not only their elements but the cultures themselves for him have meaning as elements in the overarching Culture of human history as a whole. As a result, later stages can comprehend earlier ones; and in fact, according to Hegel, a culture can be scientifically (*wissenschaftlich*) understood only when its day is done: 'The owl of Minerva spreads its wings only with the falling of the dusk'.[3]

Elements of this view have certainly been preserved in Marx's orientation to history, as the phrase quoted above demonstrates. At the same time, however, Marx's view is marked by an abandonment of the teleology essential to the Hegelian scheme. For Marx the great importance of Darwin's *On the Origin of Species* was that it dealt 'a mortal blow' to teleology in the natural sciences, while at the same time offering an empirical explanation of its 'rational meaning' – i.e. explained a teleological pattern as the result of a causal mechanism.[4] This certainly represented Marx's attitude towards teleological explanation in human history as well. He had already renounced the Hegelian vision in 1845, writing that

> History is nothing but the succession of the separate generations, each of which uses ... the productive forces handed down to it by all preceding generations, and thus, on the one hand, continues the traditional activity in completely changed circumstances and, on the other, modi-

---

3 Hegel, *Philosophy of Right* (1952 [1820]), p. 13. Ehud Sprinzak has observed that this belief of Hegel's was the ancestor of Marx's argument 'that an insightful view of society in history best to be gained only in retrospect. Classical political economy, according to this conception, can very well understand all the ... stages of history that anticipated its own, by virtue of its inclusive categories', such as those of labour, economy, etc. This explains Marx's unwillingness to write 'recipes for the cook-shops of the future': in Sprinzak's words, 'Marx's objective knowledge could not be extended to the socialist future age but had to limit itself to the critique of bourgeois society ...' (Sprinzak, 'Marx's Historical Conception of Ideology and Science' [1975], pp. 409, 410). This is why 'scientific socialism' could not be a science of the future society but only a critical understanding of the present-day movement to establish one.

4 Marx to Lasalle, 16 January 1861, in MECW, Vol. 41 (1987), p. 247.

fies the old circumstances with a completely changed activity. This can be speculatively distorted so that later history is made the goal of earlier history ... while what is designated with the words 'destiny', 'goal', 'germ', or 'idea' of earlier history is nothing more than an abstraction from later history, from the active influence which earlier history exercises on later history.[5]

Thirty-odd years later, Marx wrote one of his Russian disciples to stress that the 'historical inevitability' which *Capital* ascribed to the development of capitalism 'is expressly limited to the countries of Western Europe'.[6] Similarly, in reply to Mikhailovski, who had criticised him for a theory of 'necessary stages' in history, Marx replied that it was only his Russian critic for whom it was

> absolutely necessary to metamorphose my historical sketch of the genesis of capitalism in Western Europe into an historical-philosophical theory of general development, imposed by fate on all peoples, whatever the historical circumstances in which they are placed, in order to eventually attain this economic formation [socialism] which, with the tremendous leap of the productive forces of social labour, assures the most integral development of every individual producer. But I beg his pardon. This does me too much honor, and yet puts me to shame at the same time.

One can of course, Marx continues, come to some more or less general conclusions about history by comparing the results of detailed studies of different areas and times. But an understanding of social history will never be achieved 'by employing the all-purpose formula of a general historico-philosophical theory whose supreme virtue consists in being supra-historical'.[7]

We must conclude, then, that despite his use of Hegelian phraseology Marx did not conceive of the series 'Asiatic, ancient, feudal, and [capitalist] modes of production' as ordering all the world's cultures in one connected history. The sequence appears universal in character because capitalism, which developed from European feudalism, itself developed on the ruins of 'ancient' society, in becoming a world system transforms all the social forms it overwhelms. As

---

5   Marx and Engels, *German Ideology* (1976 [1845–46]), p. 50.
6   Marx to Zasulich, 8 March 1881, in MECW, Vol. 46 (1987), p. 71.
7   Marx to the editors of *Otechestvenniye Zapiski*, November (?), 1877, in MECW, Vol. 24 (1989), p. 199; and see Marx and Engels, *Die russische Kommune. Kritik eines Mythos* (1972), pp. 49 ff.

Marx put it in *The German Ideology*, the more capitalism expands, breaking down the isolation of hitherto relatively autonomous cultures, 'the more history becomes world history'.[8]

Even if we disregard the appearances of a teleological conception, however, can we not identify 'what deserves to be called a *theory* of history, which is not a reflective construal, from a distance, of what happens, but a contribution to understanding its inner dynamic'? That we can is the claim of G.A. Cohen in his influential book, *Karl Marx's Theory of History*, in which he attempts to reconstruct 'parts of historical materialism as a theory or infant science'.[9] Cohen's version of the theory supposedly stated in the 1859 Preface, and developed in the series of works constituting the critique of political economy, predicates of history 'a perennial tendency to productive progress, arising out of rationality and intelligence in the context of the inclemency of nature'. The basis of history is thus development of the forces of production, as measured by the ability to produce a surplus above current consumption. The basic situation of mankind is said to be one of 'scarcity', meaning that 'given men's wants and the character of external nature, they cannot satisfy their wants unless they spend the better part of their time and energy doing what they would rather not do, engaged in labor which is not experienced as an end in itself'. Human intelligence leads to the development of technologies increasingly able to meet the challenge of scarcity; human rationality in turn dictates that 'when knowledge provides the opportunity of expanding productive power [people] will tend to take it'. Productive forces exist and are utilised only within one or another set of social production relations, which therefore develop in such a way as to permit the development of productive forces. The forces – the embodiment of rationality's response to scarcity – are the primary causal element in history; 'the property of a set of productive forces which explains the nature of the economic structure embracing them is their disposition to develop within a structure of that nature'.[10]

Setting aside the difficulties of the 'functional' mode of explanation Cohen employs, this is an extremely implausible reconstruction of Marx's views, as well as an implausible understanding of history. As we have seen, Marx explicitly rejects appeals to features of the human mind, including therefore its alleged rationality, as a basis for historical explanation. At the same time, the concept of 'scarcity', drawn – like the characterisation of work as inherently

---

8   Marx and Engels, *German Ideology* (1976 [1845–46]), p. 51.

9   Cohen, *Karl Marx's Theory of History* (1978), p. 27.

10   Ibid., pp. 155, 152, 153, 161.

painful – from the ahistorical economics Marx was attacking, has no place in Marx's view. For Marx, 'wants' are precisely not 'given'; the 'creation of new needs is the first historical act'.[11] This 'creation of needs' is itself conditioned by social experience, in Marx's view; thus he held that 'the key to the riddle of the unchangeability of Asiatic societies' is supplied by 'the simplicity of the productive organism in these self-sufficing communities which constantly reproduce themselves in the same form' and 'on an unalterable division of labor'.[12] Here the nature of the social relations inhibits the progress of productivity, just as in capitalism the nature of society has led to its acceleration.

I chose Cohen to discuss because his work has been highly praised among recent contributions to the understanding of Marx. But I believe that investigation of any other attempt to develop a general theory of history from Marx's work would lead to the same result. It would otherwise be remarkable that the argument in *Capital* makes no use of a general social theory of the sort we have been discussing. We have already seen that Marx's general categories, as abstractions, are used not to generate explanations for particular phenomena but to organise the material in a comparative way. The 'guiding thread' in his theoretical fabric, as he makes clear in an interesting reference in *Capital* to the Preface to *Zur Kritik*, is that 'each particular mode of production' is the foundation of 'definite forms of social consciousness'.[13] For this reason a general theory of ideology is impossible, as we saw in Chapter 4.

It is the rejection of the possibility of general theories of society that lies behind the two-fold critique of bourgeois social thought exhibited in Marx's distinction between 'classical' and 'vulgar' economists. The former were scientific insofar as they sought to construct explanations of economic phenomena in 'the real internal framework of bourgeois relations of production', while the latter only 'flounder around in the apparent framework of those relations' –

---

11   Marx and Engels, *German Ideology* (1976 [1845–46]), p. 42. This is the basis for two important points made by Marx in his critique of political economy. First, he attacked the biological subsistence conception of wages; in contradistinction both to the classical economists and to early socialists like Lasalle, for Marx 'the number and extent of [the worker's] so-called necessary requirements, also the manner in which they are satisfied, are themselves the product of history ...' (*Capital*, Vol. I [1976 (1867)], p. 275). Second, he insisted against Malthus (and, after him, the generality of economists) that it is incorrect to speak of a natural law of population; he insisted that every form of society has its own law of population.

12   Marx, *Capital*, Vol. I (1976 [1867]), pp. 479, 478. See also Krader, *The Ethnological Notebooks of Karl Marx* (1972) and *The Asiatic Mode of Production* (1975), both strikingly missing from Cohen's bibliography, which shows little sign of acquaintance with the marxological literature.

13   Marx, *Capital*, Vol. I (1976 [1867]), pp. 175–6, n. 35. See also ibid., pp. 273 f.

apparent, that is, in 'the banal and complacent notions held by the bourgeois
agents of production about their own world, which is to them the best possible
one'.[14] Here 'the forms of appearance are reproduced directly and spontan-
eously, as current and usual modes of thought; the essential relation must first
be discovered by science'.[15] In another passage Marx invokes an analogy with
physical science: a scientific analysis of capitalism is possible only in terms of
new theoretical categories, not part of the system we are trying to explain, 'just
as the apparent motions of the heavenly bodies are intelligible only to someone
who is acquainted with their real motions, which are not perceptible to the
senses'.[16] Everyone can see that the sun rises and sets, moving around the earth;
only by looking at the solar system from a more abstract point of view – i.e.
one that sets the earth-eye view alongside those from other spatial positions –
can the 'essential relation' – the motion of the earth around the sun – be seen.
Once we know the real motion we are able to explain the apparent motion of
the sun around the earth. Similarly, Marx argued, the origin of profit, which
appears to result from the investment of capital, can be understood only from
a more abstract point of view that sees capitalism as one way of organising an
exploitative production process.

It is to be noted that for Marx neither its lack of historical consciousness nor
its theoretical inadequacies disqualified classical economics from the title of
'science'. Science is not to be identified with correct theories only, but rather
with the employment of certain standards of theory-construction and evidence in the explanation of some domain of phenomena. Yet a theory that
regards capitalist social relations as natural, inevitable forms 'can only remain
a science while the class struggle remains latent or manifests itself only in isolated and sporadic phenomena',[17] since the growth of opposition to capitalism
bears witness to the spiralling crisis cycle characteristic of the developed system, and so to the historical limit to its existence; any theory worthy of the
name will have to be able to explain this cycle and the implications of capitalism's finitude. Economics at its best was riding to an inevitable fall as a result of
its appeal, explicit or implicit, to the concept of *nature* in the analysis of social
institutions.

While classical political economy was scientific, in distinction to vulgar economics, in explaining the phenomena constituting the economic system, the
two were alike in viewing 'the capitalist order as the absolute and ultimate

14   Ibid., pp. 174–5, n. 34.
15   Ibid., p. 682.
16   Ibid., p. 433.
17   Ibid., p. 96.

form of social production, instead of as a historically transient stage of development'.[18] Since capitalism does not in fact represent the emergence to clear view
of social processes and institutions essential to any and every form of society,
this viewpoint was bound to lead to errors and an inability to explain fundamental characteristics of capitalism, such as the tendency toward depression
and crisis. Thus 'classical political economy's ... uncritical acceptance of the
categories [of capitalist society] led it into inextricable confusions and contradictions ..., while it offered a secure base of operations to the vulgar economists
who, in their shallowness, make it a principle to worship appearances only'.[19]

By revealing the natural laws of the production and distribution of wealth
political economy was supposed to clarify the range of choices open to social
policymakers. Outside these choices there was nothing.

> The belated scientific discovery that the products of labour, in so far as
> they are values, are merely the material expressions of the human labour
> expended to produce them, marks an epoch in the history of mankind's
> development, but by no means banishes the semblance of objectivity pos-
> sessed by the social characteristics of labour. Something which is only
> valid for this particular form of production ... appears to those caught up
> in the relations of commodity production (and this is true both before and
> after the above-mentioned scientific discovery) to be just as ultimately
> valid as the fact that the scientific dissection of the air into its component
> parts left the atmosphere itself unaltered in its physical configuration.[20]

Just as the chemical analysis of the atmosphere did not alter but only explained
the composition of the air, so the classical analysis of value in terms of social
labour time did not alter the reality thus described. This is of course exactly
what we should expect from a scientific analysis. However, while Marx's explanation of *why* labour time appears in the form of value also does not alter the
reality explained, it does suggest the possibility of such change in a way the
classical analysis does not. By describing 'value' as a category of social behaviour, as a concept in terms of which goods are *dealt with*, Marx shows that it
is not just descriptive but also constitutive of the social system. While natural
properties of goods exist independently of social action, a property like value
that is itself a product of social behaviour can cease to exist even as it was once
brought into existence.

---

18   Ibid.
19   Ibid., p. 679.
20   Ibid., p. 167.

On the one hand, Marx criticises the portrayal of the institutions of modern society, by economists and by social theorists generally, as natural givens. He emphasises, in contrast, not only the transitory, historical character of social structures but also the fact that they are products of human activity rather than of a 'logic of history'. On the other hand, Marx describes his own work on the pattern of natural science, comparing it explicitly to the physicists' research into natural processes.[21] This paradox reflects, not a self-contradiction in Marx's understanding of his enterprise, but the nature of the object of enquiry – capitalism – itself. It is a feature of this social system that the consequences of people's continuing adherence to the rules and institutions defining it affect them as forces over which they have no control, and which indeed the imaginative effort of science is necessary to understand. Truly, says Marx, the determination of value by social labour time 'asserts itself as a regulative law of nature', but – quoting his friend Engels – 'What are we to think of a law which can only assert itself through periodic crises? It is just a natural law which depends on the lack of awareness of the people who undergo it'.[22] As we have seen, the law would remain in effect even if people did understand it: beyond understanding, the world requires actual changing. But it would no longer appear *natural*.

Thus Marx suggests that social science comes into existence when society appears as a force (or structure) dominating the experience of those who make it up. In particular, Marx suggests that it is capitalism itself which creates the conditions for a science of society. It is certainly the case, despite the work of Plato, Aristotle, and Ibn Kaldun, that the social sciences as we know them have their origin in the seventeenth century. Joseph Schumpeter attempted an explanation of this historical fact:

> [C]ommonsense knowledge, relative to scientific knowledge, goes much farther in the economic field than it does in almost any other. It is perfectly understandable, therefore, that economic questions, however important, took much longer in eliciting specifically scientific curiosity than did natural phenomena. Nature harbors secrets into which it is exciting to probe; economic life is the sum total of the most common and most drab experiences.[23]

---

21   See ibid., p. 90.

22   Ibid., p. 168 n. 30.

23   Schumpeter, *History of Economic Analysis* (1954), p. 53.

It obviously did not occur to him that the subject matter of economics – the capitalist market economy – not only was not banal but did not exist until the modern era. As Marx said,

> The value character of the products of labour becomes firmly established only when they act as magnitudes of value. These magnitudes vary continually, independently of the will, foreknowledge, and actions of the exchangers. Their own movement within society has for them the form of a movement made by things.[24]

It was this actual experience of a world of autonomous prices, an historical novelty, which suggested the need for a science of this new sphere of experience, the market economy. Given the rapid rise of the language of natural science to its position as privileged mode of discourse, reflecting both the break with the theological ideology of the *ancien régime* and the technological needs of industrial capitalism, the extension of scientific thinking to the study of society – in Hume's words, 'An attempt to introduce the experimental Method of Reasoning into Moral subjects'[25] – was inevitable.

Indeed, it was first in the eighteenth century that the word 'society' took on its modern meaning as the body of institutions and relationships, characterised by special laws, in which the individual finds himself.

> It is not until the 18th century, in 'bourgeois society', that the various forms of the social nexus confront the individual as merely a means towards his private ends, as external necessity. But the epoch which produces this standpoint, that of the isolated individual, is precisely the epoch of the hitherto most highly developed social ... relations.[26]

Hence the subjectivist strain in social science, seeing the essence of social life in individual consciousness; hence also the problem of explaining how the interaction of many individuals produces results affecting each person as the working of an Invisible Hand, and the conception of social processes as a domain of 'social facts'. It seems to follow from this understanding of the social sciences as products of capitalism that socialist revolution can be expected to

---

24   Marx, *Capital*, Vol. I (1976 [1867]), p. 167.
25   [Thus the eighteenth-century philosopher described his *Treatise of Human Nature* (1739–40).]
26   Marx, 'Economic Manuscripts of 1857–58', in MECW, Vol. 28 (1986), p. 18. See the article on 'Society', in Williams, *Keywords* (1976), pp. 243–7.

bring with it, along with the abolition of wage labour and the state, the disappearance of the subject matter of economics and political science – more
generally, in G.A. Cohen's phrase, 'the withering away of social science'. This
is already suggested in Marx's comparison of the mystery of commodity relations with the absence of economic facts in other forms of society. Thus in
medieval Europe, 'precisely because relations of personal dependence form the
given social foundation, there is no need for labour and its products to assume
a fantastic form different from their reality'. In this society people nevertheless
experienced social reality ideologically, in terms of such concepts as kinship,
honour, loyalty, and the divine will – 'the motley feudal ties that bound men
to their "natural superiors"' that capitalism (as Marx put it in the *Commun-
ist Manifesto*) 'has ruthlessly torn asunder'. On the other hand, in the future
socialist society as Marx imagined it, 'an association of free men, working with
the means of production held in common, and expressing their many different forms of labour power in full self-awareness as one single labour force', the
relations of the producers with each other, their labour, and its products, would
again be 'transparent'. The gap between appearance and reality filled today by
economic science would no longer exist.[27] Marx believed that a tendency in
this direction was to be seen even within capitalism, as in the course of the
class struggle the workers discover the historically limited character of the laws
of economics, so that the continued effectiveness of these laws becomes a matter of political action.

> Thus as soon as the workers learn the secret of why it happens that the
> more they work, the more alien wealth they produce, and that the more
> the productivity of their labour increases, the more does their very func-
> tion as a means for the valorization of capital become precarious; … as
> soon as, by setting up trade unions, etc., they try to organize planned co-
> operation between the employed and the unemployed in order to obviate
> or to weaken the ruinous effects of [the] natural law of capitalist pro-
> duction on their class, so soon does capital and its sycophant, political
> economy, cry out at the infringement of the 'eternal' and so to speak 'sac-
> red' law of supply and demand. Every combination between employed
> and unemployed disturbs the 'pure' action of this law.[28]

---

27   Marx, *Capital*, Vol. I (1976 [1867]), pp. 170, 171.
28   Ibid., pp. 793–4. Marx goes on to point out that the bourgeoisie are quite happy to violate
      the 'sacred' law of supply and demand themselves, in the case of labour shortages.

With the triumph of communism, furthermore, Marxian theory itself would cease to have meaning. Insofar as the aim of this theory is the unmasking of the social reality beneath the appearances of capitalist society, it would have no place in a world in which 'the practical relations of everyday life between man and man, and man and nature, generally present themselves to him in a transparent and rational form'.[29] Marxian science, in other words, will of necessity share the fate of its bourgeois antagonist.

This would not mean an end of scientific thinking about society, but only that of bourgeois ideology and its critique. The disappearance of those phenomena rooted in the peculiarities of capitalism would not bring a sudden solution of all the mysteries of social behaviour: the nature of grammar and its relation to spoken language, for instance, or the processes by which an infant becomes a human being would still remain invisible to the non-scientific eye. In addition, the attempt to construct a democratically and consciously regulated society would require exactly that questioning of assumptions and procedures which plays an essential role in scientific theorising. To be scientific here would mean consciously to attempt to place existing and proposed institutions and procedures in the context of the past, on the one hand, and in that of the imagined and desired future, on the other. The goal of democratic decision-making would require attention paid to comparison and mediation among the variety of cultural forms constituting society on the worldwide and local levels. But one imagines a flavour here quite different from that of today's social science, as the aim would be explicitly not so much the discovery of necessities as the expansion of the realm of choice.

Such a future implies a transformation in the conception of social knowledge which goes far beyond the discarding of the illusions of the social pseudo-sciences. Two aspects of this transformation – the abandonment of 'value-freedom' as a scientific ideal, and an alteration of the idea of science itself – may be discussed here, however briefly, as they may be discerned in trends of the present time, bound up with the continuing development of the existing social system. The two are intimately connected in the complex idea of 'nature' and of science as the 'objective' study of it. Basic to the concept of 'nature' involved here is the idea of a 'natural order', represented by a system of laws, existing independently of the human observer, that it is the aim of science to discover. This is a world of 'objective' facts, as opposed to the ('subjective') conventions and values of human invention.

---

29   Ibid., p. 173.

It is in accordance with this set of concepts that social science is taken to be the study of human nature, or of the nature of society, construed as a set of drives, instincts, innate mental structures, or social necessities in terms of which complex systems can be explained. A classic version of the ideology of modern science imagined a unified science, in which sociology, individual psychology, biology, chemistry, and physics study the properties of the same basic stuff at varying degrees of complexity. But the relevant concept of nature may be shared by those, like Durkheim, who think in terms of irreducibly social facts.

Astronomer Brun van Albada suggests that not just social science but this idea of 'science' itself is a part of the development of capitalism, 'born of the presumption of that class which had the illusion of having solved all social problems and believed it had the power to dominate nature as it dominated society. Thanks to the bourgeoisie the world was now in order, and in good order, in harmony with the demands of reason. It ought to run regularly, like a well-designed and well-oiled mechanism'.[30] Van Albada continues by pointing out that the shocks to the bourgeoisie's sense of cosmic order and its place in it occasioned by the economic and social upheavals of the latter half of the nineteenth century were reflected in the displacement of the confident materialism of the eighteenth century by 'a semi-idealist conception ... which, without entirely denying the existence of the material world, could assign to man only the role of powerless and fearful observer',[31] a conception exemplified by Mach's philosophy of nature. Certainly the thirty years' renewed expansion of capitalism that followed World War II were accompanied by a renewed faith in the power of science to understand and control the world, social and natural – this was the age both of the atom bomb and of 'macroeconomic fine-tuning'. In this context the dominance of positivist philosophy of science is not surprising. But the optimism of capitalism's earlier periods could not survive the century of total war, and the social relativism produced in part by the loss of faith in progress as of the essence of capitalism and in part by the apparently stable confrontation with state-run societies has made possible the development of 'post-modern' (if not yet post-capitalist) philosophies of science.[32]

---

30   Albada, 'Sciences de la nature et société' (1969).
31   Ibid.
32   The phrase is borrowed from Stephen Toulmin's article, 'The Construal of Reality: Criticism in Modern and Postmodern Science' (1982), expressing a view on the relation of natural to social science somewhat similar to my own.

One consequence of this intellectual development was a change in the understanding of laws of nature. The concept of 'natural law', applied for most of its history both to the investigation of physical nature and the evaluative study of social institutions, is of course itself a metaphor taken from social reality. The medieval Christian picture of God as law-giving sovereign of creation gradually gave way to the image of God as engineer, designer of the world machine. The disappearance of the divine rule-setter from today's science has left the metaphor as designation of a form of description, or as a means to the explanation, of natural phenomena.

We are coming to see that just as what seemed to be social givens are not universal, so truth need not be understood as something given and waiting to be discovered, but can be construed as something historically and socially constructed, without for all that becoming arbitrary or 'subjective'. The concept of 'natural order', similarly, can now be seen to draw its meaning from the fact that we are able to construct theories ordering and explaining our experience, just as it is only by our construction of categories and theory-like systems of thinking that our experience has shape at all. We now have learned that just as on the individual psychological level there is no perception without representation, so scientific experiments are meaningful only in the context of sets of theoretical constructs.

Science, in a word, is being perceived more and more on the model of a product of social labour, both as a final product and as a tool for future production, although few are as clear about this as is J.R. Ravetz in his description of scientific work as a type of 'craft' exercised on 'artificially constructed objects' (intellectual constructs) and his conclusion that 'the special character of achieved scientific knowledge' – its justifiable claim to truth – 'is explained by the complex social processes of selection and transformation of the results of research'.[33] This new understanding of the place of science in the social labour process has been no doubt in part suggested by the actual social position of the scientist, far different today from that of the eighteenth or even the nineteenth century. In the age of 'big science', involving massive state and private investment, the scientist can no longer be seen as a disinterested and independent seeker of truth. This situation has led to a change

> as radical as that which occurred ... when independent artisan producers were displaced by capital-intensive factory production employing hired labor ... With his loss of independence, the scientist falls into one of these

---

33   Ravetz, *Scientific Knowledge and its Social Problems* (1971), p. 72.

roles: either an employee, working under the control of a superior; or an individual outworker for investing agencies, existing on a succession of small grants; or he may be a contractor, managing a unit or establishment which produces research on a large scale by contract with agencies.[34]

One casualty of this situation is the image of the scientist as an individual mind seeking truth through analysis of its experience of the 'external world'. The alternative is a conception of scientific truth not in terms of a correspondence of mental with natural objects or structures, but as socially communicable and confirmable knowledge.

This conception of science also erodes the distinction between science as search for truth, on the one hand, and its technical application, a non-scientific matter to be decided by reference to values, on the other. The impossibility of sustaining this distinction in the twentieth century was perhaps demonstrated most clearly by Enrico Fermi's [purported] use of it in explaining his feelings about the atomic bombing of Hiroshima: while unfortunate from the human point of view, he explained, it was a wonderful piece of physics.[35] The classical formulation of this ideal for the social sciences is that of Max Weber, in his addresses on science and politics as vocations. For Weber the 'scientific attitude' implied freedom from the evaluative stance essential to political activity. This attitude reproduces within the social sciences the opposition between the (naturally) given and (culturally) chosen or created aspects of society, as with the difference between the laws of economics and policy

---

34   Ibid., p. 44. Ravetz quotes the Soviet physicist Kapitsa:
      The year that Rutherford died (1938) there disappeared forever the happy days of free
      scientific work which gave us such delight in our youth. Science has lost her free-
      dom. Science has become a productive force. She has become rich but she has become
      enslaved and part of her is veiled in secrecy. I do not know whether Rutherford would
      continue nowadays to joke and laugh as he used to do (ibid., p. 32).

35   That such attitudes were hardly peculiar to Fermi is shown by other memoirs of atomic
      physicists, as well as by the continued work of some hundred thousand scientists on
      weapons research. In the words of Jacob Bronowski, in his book *Science and Human Values*
      [*sic*], 'Science has nothing to be ashamed of even in the ruins of Nagasaki. The shame is
      theirs who appeal to other than the human imaginative values that science has evolved'
      (1961, p. 83, cited by Ravetz, *Scientific Knowledge* [1971], p. 65, n. 41) – values which, it
      need hardly be pointed out, did nothing to prevent the bomb-designers from proceed-
      ing with their work. [Jeffrey Lewis has cast serious doubt on Fermi's ever having made the
      remark credited to him; see 'The Thing is Superb Physics', https://www.armscontrolwonk
      .com/archive/203509/the-thing-is-superb-physics (last accessed 25/1/2018). But the gen-
      eral point, alas, remains.]

choices evoked by the title of Arthur Okun's book, *Equality and Efficiency: The Big Tradeoff*.[36]

The same attitude appeared within Marxism, as exemplified by Eduard Bernstein's distinction between socialism as an ethical ideal and Marxism as a scientific theory of the capitalist economy.[37] For this Bernstein seemingly could claim ancestry in Marx's characterisation of his own viewpoint as one 'from which the development of the economic formation of society is viewed as a process of natural history'. Marx praised the English factory inspectors for their freedom 'from partisanship and respect of persons', and observed, in anticipation of *Capital*'s critical reception, that

> In the domain of political economy, free scientific inquiry does not merely meet the same enemies as in all other domains. The peculiar nature of the material it deals with summons into the fray on the opposing side the most violent, sordid and malignant passions of the human breast, the Furies of private interest.

In the face of this, he welcomed 'every opinion based on scientific criticism'.[38]

Yet Marx had no wish to pretend that his theoretical labours were 'value-free'. His aim was to serve what he considered the practical needs of the working class in its struggle against capitalism. But this for Marx meant not abandonment of the claim to scientific truth, but the opposite. Those who wish to control their social (as their natural) conditions of life need to understand the situations in which they find themselves and the possible choices of action within these situations. In fact, it seems clear that Marx anticipated the picture of science which only now is gaining more general acceptance as the inadequacy of the older picture becomes inescapable. With the supposed conflict between objectivity and values Marx abandoned both the idea that there is an eternal social order to be discovered and the scientist's claim to a special position of social expertise, based on his supposed character of a seeker after truth, above the passions of the laity.

Thus by 'scientific socialism', as Marx put it in reply to criticism by Bakunin, he meant – in opposition to 'utopian socialism which seeks to foist new

---

36   Okun, *Equality and Efficiency* (1975).

37   See Eduard Bernstein, *Wie ist wissenschaftlicher Sozialismus möglich?* (1901). Bernstein shared Max Weber's neo-Kantianism. Not too much should be made of this, however, for our purposes here, as this philosophical orientation, like Kant's itself, was only a particular manifestation of a general cultural attitude.

38   Marx, *Capital*, Vol. I (1976 [1867]), pp. 91, 92, 93.

fantasies upon the people' – 'the comprehension of the social movement created by the people themselves'.[39] The historical process Marx was interested
in, as a socialist theoretician, would consist precisely in people's attempts to
change the society in which they find themselves, just as that society's continued existence consists in their continued obeisance to the laws of economics.
Scientific work, in leading to a better understanding of society and so of the
tasks involved in changing it, should serve as an element of these attempts.

'The production of ideas, of conceptions, of consciousness', Marx says in *The
German Ideology*,

> is at first directly interwoven with the material activity and the material
> intercourse of men – the language of real life ... The same applies to men-
> tal production as expressed in the language of the politics, laws, morality,
> religion, metaphysics, etc. of a people. Men are the producers of their con-
> ceptions, ideas, etc., that is, real, active men, as they are conditioned by
> a definite development of their productive forces and of the intercourse
> corresponding to these ... Consciousness [*das Bewusstsein*] can never be
> anything else than conscious being [*das bewusste Sein*], and the being of
> men is their actual life-process.[40]

Once consciousness is construed as the organisation of human activity, then
revolutionary consciousness, like its opposite, can be understood as the systems and quasi-systems of conceptions by means of which people organise
their revolutionary (or non-revolutionary) behaviour. Like everyone, revolu-
tionaries think about what they are doing: the theoreticians among them are
those who try to explore, systematise, and explain the social interactions that
constitute the social status quo and the movement against it.

This attitude was reflected in Marx's conception of the tasks of intellectuals in the socialist movement. He put his writing skills at the service of the
First International, preparing position statements, official communications,
and so forth. In addition, we should note the project of an *Enquête Ouvrière*,
a questionnaire which Marx published in the Parisian *Revue Socialiste* in 1880,
and had reprinted and distributed to workers' groups, socialist and democratic
circles, 'and to anyone else who asked for it' in France. The text has the form of
101 questions, about working conditions, wages, hours, and effects of the trade
cycle, and also about workers' defence organisations, strikes and other forms

---

39   Marx, 'Notes on Bakunin', in Marx and Engels, *Werke*, Vol. 18, translation quoted from Hunt,
     The Political Ideas of Marx and Engels, Vol. 1 (1974), p. 326.
40   Marx and Engels, *The German Ideology* (1976 [1845–46]), p. 36.

of struggle, and their results. Though this might be described as the first soci-ological survey, its preface urges workers to reply, not to meet the data needs of sociologists or economists, but because only workers can describe 'with full knowledge the evils which they endure', just as 'they, and not any providen-tial saviors, can energetically administer the remedies for the social ills from which they suffer'. The role which Marx intended to take on was that of the col-lection, organisation, and transmission of this information; thus, the results of the *Enquête* were to be analysed in a series of articles for the *Revue* and, even-tually, a book.[41]

The main task which Marx set himself as a revolutionary intellectual, how-ever, was the task of theory: the elaboration of a set of concepts, at a fairly abstract level, that would permit a better comprehension of the struggle be-tween labour and capital. He prefaced the French serial edition of the first volume of *Capital* with an expression of pleasure, that 'in this form the book will be more accessible to the working class – a consideration which to me out-weighs everything else'.[42] The function of theory was to help the movement as a whole clarify its problems and possibilities; it did not, in Marx's view, place the theorist in a dominating (or 'hegemonic', as the currently fashionable euphem-ism has it) position vis-à-vis the movement, but was rather what he had to contribute to a collective effort.

That effort, however, was to be abandoned. *Capital* was transformed by the history of the Second and Third Internationals into the 'Bible of the work-ing class' – an ironic fate for the product of Marx's scientific efforts. Today, in the absence of a working-class movement, Marxism has become a specialisa-tion for intellectuals, and Marx's critique of the ideology of capitalist society grist for the social-scientific mills that continue to produce that ideology. Basic elements of Marx's approach to the analysis of social action and conscious-ness have increasingly come to be adopted by the social sciences – strikingly enough, with the greatest success and the least resistance at the points farthest from the critique of present-day reality, such as the study of classical antiquity or the anthropology of 'simpler societies'. But in their inability to understand the culture that gave birth to them, the social sciences, and in particular eco-nomics, have not advanced much since Marx's day.

Commenting on a recent book on this state of affairs, a reviewer noted the author's

---

41   Marx, *Selected Writings in Sociology and Social Philosophy* (1963), pp. 210–11.
42   Marx, *Capital*, Vol. I (1976 [1867]), p. 104.

firmly held belief that the notorious absence of a universally accepted conceptual core in social science, and of commonly agreed standards by which such core-concepts could be selected, has been engendered in no small measure by confusion as to the exact epistemological status of social-scientific concepts; and that the resulting disarray could be largely rectified, if only social scientists were to embrace the right philosophy of science and recognize the true nature of scientific concepts.[43]

Despite my own sympathy for anything suggesting the social utility of employment of philosophers, the weakness of this suggestion is obvious. What needs explanation is the persistent failure, throughout the history of these 'folk sciences' (as Ravetz calls them), to embrace conceptions of society and of science permitting real progress in the direction of social knowledge. I have hoped to defend the powers of scientific rationality against barriers supposedly discovered through philosophical analysis. But my wish is not so much for the improvement of the social sciences as academic disciplines as for the imaginative rejection of the conditions of present-day society by their victims. It is this refusal of the laws of capitalist reality which will make possible both their understanding and their abolition.

---

43   Bauman, 'Review of William Outhwaite, *Concept Formation in Social Science*' (1983), p. 441.

References

Albada, Brun van [Gale Bruno van Albada] 1969, 'Sciences de la nature et société', *Les Cahiers du communisme de conseils*, 3.

Barden, Garret 1972, 'Method and Meaning', in *Zande Themes*, edited by Andre Singer and Brian V. Street, Totowa, NJ: Rowman and Littlefield.

Barnes, Barry 1974, *Scientific Knowledge and Sociological Theory*, London: Routledge and Kegan Paul.

Bauman, Zygmunt 1983, 'Review of William Outhwaite, *Concept Formation in Social Science*', *Times Literary Supplement*, 29 April.

Beller, Mara 1999, *Quantum Dialogue: The Making of a Revolution*, Chicago: University of Chicago Press.

Bernstein, Eduard 1901, *Wie ist wissenschaftlicher Sozialismus möglich?*, Berlin: Sozialistische Monatshefte.

Bernstein, Richard J. 1978, *The Restructuring of Social and Political Theory*, Philadelphia: University of Pennsylvania Press.

Bottomore, Tom 1971, *Sociology: A Guide to Problems and Literature*, New York: Pantheon.

Bourdieu, Pierre 2004, *Science of Science and Reflexivity*, trans. Richard Nice, Chicago: University of Chicago Press.

Bouritsas, Leonidas 1988, 'Book Review: *Social Knowledge*', *Review of Radical Political Economy*, 20, no. 1: 128–30.

Brodbeck, May (ed.) 1968, *Readings in the Philosophy of the Social Sciences*, New York: Macmillan.

Bronowski, Jacob 1961, *Science and Human Values*, London: Hutchinson.

Canterbery, E. Ray and R.J. Burkhardt 1979, 'Economics: The Embarrassed Science', *National Science Foundation EVIST Research Report*.

Clark, Stuart 1980, 'Inversion, Misrule, and the Meaning of Witchcraft', *Past and Present*, 87: 98–127.

Cohen, G.A. 1978, *Karl Marx's Theory of History*, Princeton: Princeton University Press.

Davidson, Donald 1963, 'Actions, Reasons, and Causes', *Journal of Philosophy*, 60, no. 23: 685–700.

Easlea, Brian 1980, *Witch Hunting, Magic, and the New Philosophy: An Introduction to the Debates of the Scientific Revolution, 1450–1750*, Sussex: The Harvester Press.

Eichner, Alfred S. (ed.) 1983, *Why Economics is Not Yet a Science*, Armonk: M.E. Sharpe.

Evans-Pritchard, E.E. 1965, 'The Comparative Method in Social Anthropology', in *The Position of Women in Primitive Societies and Other Essays in Social Anthropology*, New York: Free Press.

Evans-Pritchard, E.E. 1976, *Theories of Primitive Religion*, Oxford: Oxford University Press.

Evans-Pritchard, E.E. 1976, *Witchcraft, Oracles, and Magic among the Azande*, Oxford: Oxford University Press.

Feyerabend, Paul K. 1962, 'Explanation, Reduction, and Empiricism', in *Minnesota Studies in the Philosophy of Science*, Vol. III, edited by Herbert Feigl and Grover Maxwell, Minneapolis: University of Minnesota Press.

Fisher, Lawrence E. and Oswald Werner 1978, 'Explaining Explanation: Tension in American Anthropology', *Journal of Anthropological Research*, 34, no. 2: 194–218.

Flathman, Richard E. 2000, 'Wittgenstein and the Social Sciences: Critical Reflections Concerning Peter Winch's Interpretations and Appropriations of Wittgenstein's Thought', *History of the Human Sciences*, 13, no. 2: 1–15.

Galison, Peter 1997, *Image and Logic: A Material Culture of Microphysics*, Chicago: University of Chicago Press.

Garfinkel, Alan 1981, *Forms of Explanation: Rethinking the Questions of Social Theory*, New Haven: Yale University Press.

Geertz, Clifford 1964, 'Ideology as a Cultural System', in *Ideology and Discontent*, edited by David E. Apter, New York: Free Press.

Gellner, Ernest 1977, 'Concepts and Society', in *Rationality*, edited by Bryan R. Wilson, Oxford: Blackwell.

Gerth, H.H. and C. Wright Mills (eds) 1946, *From Max Weber: Essays in Sociology*, New York: Oxford University Press.

Goody, Jack 1969, *Comparative Studies in Kinship*, Stanford: Stanford University Press.

Hacking, Ian 1983, *Representing and Intervening: Introductory Topics in the Philosophy of Natural Science*, Cambridge: Cambridge University Press.

Hallen, Barry and J. Olubi Sodipo 1997 [1986], *Knowledge, Belief, and Witchcraft: Analytic Experiments in African Philosophy*, Palo Alto: Stanford University Press.

Harris, Marvin 1979, *Cultural Materialism: The Struggle for a Science of Culture*, New York: Random House.

Harris, Zellig et al. 1989, *The Form of Information in Science: Analysis of an Immunology Sublanguage*, Dordrecht: Kluwer.

Hayek, Friedrich A. 1955, *The Counter-Revolution of Science*, New York: Free Press.

Hegel, G.W.F. 1952 [1820], *Philosophy of Right*, trans. T.M. Knox, Oxford: Oxford University Press.

Hempel, Carl 1950, 'Problems and Changes in the Empiricist Criterion of Meaning', *Revue Internationale de Philosophie*, 11: 41–63.

Hollis, Martin 1973, 'Reason and Ritual', in *The Philosophy of Social Explanation*, edited by Alan Ryan, Oxford: Oxford University Press.

Horton, Robin 1977, 'African Traditional Thought and Western Science', in *Rationality*, edited by Bryan Wilson, Oxford: Blackwell.

Hunt, Richard N. 1974, *The Political Ideas of Marx and Engels*, Volume I, Pittsburgh: University of Pittsburgh Press.

Hutten, Ernest H. 1962, *The Ideas of Physics*, Edinburgh: Oliver & Boyd.

Joseph, Geoffrey 1980, 'The Many Sciences and the One World', *Journal of Philosophy*, 77, no. 11: 773–91.

Kluckhohn, Clyde 1962, *Culture and Behavior*, New York: Free Press.

Knoop, Todd A. 2004, *Recessions and Depressions: Understanding Business Cycles*, Westport, CT: Praeger.

Knorr-Cetina, Karin 1981, *The Manufacture of Knowledge: An Essay on the Constructivist and Contextual Nature of Science*, Oxford: Pergamon.

Koopmans, Tjalling 1968, 'The Construction of Economic Knowledge', in *Readings in the Philosophy of the Social Sciences*, edited by May Brodbeck, New York: Macmillan.

Korsch, Karl 1963 [1938], *Karl Marx*, New York: Russell and Russell.

Krader, Lawrence 1972, *The Ethnological Notebooks of Karl Marx*, Assen: van Gorcum.

Krader, Lawrence 1975, *The Asiatic Mode of Production*, Assen: van Gorcum.

Kuhn, Thomas S. 1962, *The Structure of Scientific Revolutions*, Chicago: University of Chicago Press.

Latour, Bruno and Steve Woolgar 1979, *Laboratory Life: The Social Construction of Scientific Facts*, Beverly Hills: Sage Publications.

Lepenies, Wolf 1981, 'Anthropological Perspectives in the Sociology of Science', in Everett Mendelsohn and Yehuda Elkana (eds), *Sciences and Cultures: Anthropological and Historical Studies of the Sciences*, Dordrecht: Reidel.

Lloyd, G.E.R. 1979, *Magic, Reason, and Experience: Studies in the Origin and Development of Greek Science*, Cambridge: Cambridge University Press.

Lukes, Stephen 1977, 'Some Problems about Rationality', in *Rationality*, edited by Bryan R. Wilson, Oxford: Blackwell.

Machlup, Fritz 1969, 'If Matter Could Talk', in *Philosophy, Science, and Method: Essays in Honor of Ernest Nagel*, edited by Sidney Morgenbesser et al., New York: St Martin's.

Mackie, J.L. 1975, 'Ideological Explanation', in *Explanation*, edited by Stephen Korner, New Haven: Yale University Press.

MacIntyre, Alasdair 1962, 'A Mistake about Causality in Social Science', in *Philosophy, Politics and Society* (second series), edited by P. Laslett and W.G. Runciman, Oxford: Blackwell.

MacIntyre, Alasdair 1973, 'The Idea of a Social Science', in *The Philosophy of Social Explanation*, edited by Alan Ryan, Oxford: Oxford University Press.

MacIntyre, Alasdair 1977, 'Is Understanding Religion Compatible with Believing?', in *Rationality*, edited by Bryan R. Wilson, Oxford: Blackwell.

Malinowski, Bronislaw 1927, *Sex and Repression in Savage Society*, London: Routledge.

Malinowski, Bronislaw 1932, *The Sexual Life of Savages in North-West Melanesia*, New York: Harcourt, Brace & Jovanovich.

Mandelbaum, Maurice 1969, 'Functionalism in Social Anthropology', in *Philosophy, Science, and Method: Essays in Honor of Ernest Nagel*, edited by Sidney Morgenbesser et al., New York: St Martin's.

Mandelbaum, Maurice 1977, *The Anatomy of Historical Knowledge*, Baltimore: Johns Hopkins Press.

Marx, Karl 1963, *Selected Writings in Sociology and Social Philosophy*, Harmondsworth: Penguin.

Marx, Karl 1976 [1867], *Capital*, Volume I, Harmondsworth: Penguin.

Marx, Karl 1978 [1885], *Capital*, Volume II, Harmondsworth: Penguin.

Marx, Karl 1981 [1894], *Capital*, Volume III, Harmondsworth: Penguin.

Marx, Karl 1987 [1859], *A Contribution to the Critique of Political Economy*, in Karl Marx and Frederick Engels, *Collected Works*, Volume 29, New York: International Publishers.

Marx, Karl 1986–87 [1857–58], 'Economic Manuscripts of 1857–58' [*Grundrisse*], in Karl Marx and Frederick Engels, *Collected Works*, Volumes 28 and 29, New York: International Publishers.

Marx, Karl 1976 [1863–66], 'Results of the Immediate Process of Production', in Karl Marx, *Capital*, Volume I, Harmondsworth: Penguin.

Marx, Karl 1988–91 [1861–63], 'Economic Manuscript of 1861–63' [*Theories of Surplus-Value*], in Karl Marx and Frederick Engels, *Collected Works*, Volumes 30–33, New York: International Publishers.

Marx, Karl and Friedrich Engels 1972, *Die russische Kommune. Kritik eines Mythos*, Munich: Carl Hanser.

Marx, Karl and Frederick Engels 1976 [1845–46], *The German Ideology*, in Karl Marx and Frederick Engels, *Collected Works*, Volume 5, New York: International Publishers.

Mattick, Paul 1981, 'Some Aspects of the Value-Price Problem', *Économies et Sociétés* (Cahiers de l'ISMEA, Série S) 15, nos. 6 and 7: 725–81.

Mattick, Paul 2003, *Art in Its Time: Theories and Practices of Modern Aesthetics*, London: Routledge.

Mattick, Paul 2018, *Theory as Critique: Essays on* Capital, Leiden: Brill.

McLeod, Malcolm D. 1972, 'Oracles and Accusations among the Azande', in *Zande Themes*, edited by Andre Singer and Brian V. Street, Totowa, NJ: Rowman and Littlefield.

Morf, Otto 1970, *Geschichte und Dialektik in der politischen Ökonomie*, Frankfurt am Main: Europäische Verlagsanstalt.

Morgenstern, Oskar 1963, *On the Accuracy of Economic Observations*, Princeton: Princeton University Press.

Morgenstern, Oskar 1972, 'Thirteen Critical Points in Contemporary Economic Theory', *Journal of Economic Literature*, 10, no. 4: 1163–89.

Nadel, S.F. 1953, *The Foundations of Social Anthropology*, Glencoe: Free Press.

Nagel, Ernest 1963, 'On the Method of Verstehen as the Sole Method of Philosophy', in *Philosophy of the Social Sciences*, edited by Maurice Natanson, New York: Random House.

Nagel, Ernest 1961, *The Structure of Science*, New York: Harcourt, Brace & World.

Natanson, Maurice 1963, 'A Study in Philosophy and the Social Sciences', in *Philosophy of the Social Sciences*, edited by Maurice Natanson, New York: Random House.

Newton, Isaac 1966, *Principia Mathematica*, trans. A. Motte, revised F. Cajori, Los Angeles: University of California Press.

Nowak, Leszek 1978, 'Weber's Ideal Types and Marx's Abstraction', in *Neue Hefte für Philosophie 13: Marx' Methodologie*, Göttingen: Vandenhoek & Ruprecht.

Okun, Arthur M. 1975, *Equality and Efficiency: The Big Tradeoff*, Washington, DC: The Brookings Institution.

Piaget, Jean and Barbel Inhelder 1969, *The Psychology of the Child*, New York: Basic Books.

Pickering, Andy 1984, *Constructing Quarks: A Sociological History of Particle Physics*, Chicago: University of Chicago Press.

Pickering, Andy 1990, 'Knowledge, Practice, and Mere Construction', *Social Studies of Science*, 20, no. 4: 682–729.

Pitt-Rivers, Julian 1971, 'On the Word "Caste"', in *The Translation of Culture: Essays to E.E. Evans-Pritchard*, edited by T.O. Beidelman, London: Tavistock Publications.

Quine, W.V.O. 1951, 'Two Dogmas of Empiricism', *Philosophical Review*, 60: 20–43.

Quine, W.V.O. 1960, *Word and Object*, Cambridge, MA: MIT Press.

Radcliffe-Brown, Alfred 1952, *Structure and Function in Primitive Society*, New York: The Free Press.

Ravetz, J.R. 1971, *Scientific Knowledge and Its Social Problems*, Oxford: Oxford University Press.

Routh, Guy 1975, *The Origin of Economic Ideas*, White Plains: International Arts and Sciences Press.

Ruddick, Sara 1969, 'Extreme Relativism', in *Language and Philosophy*, edited by Sidney Hook, New York: New York University Press.

Sarana, Gopala 1975, *The Methodology of Anthropological Comparisons*, Tucson: University of Arizona Press.

Schumpeter, Joseph A. 1954, *History of Economic Analysis*, New York: Oxford University Press.

Schutz, Alfred 1973, *The Phenomenology of the Social World*, Evanston: Northwestern University Press.

Seligman, Brenda Z. and C.G. Seligman 1932, *Pagan Tribes of the Nilotic Sudan*, London: Routledge.

Shapere, Dudley 1982, 'The Concept of Observation in Science and Philosophy', *Philosophy of Science*, 49, no. 4: 485–525.

Shoeffler, Sidney 1955, *The Failures of Economics: A Diagnostic Study*, Cambridge: Cambridge University Press.

Silk, Leonard 1983, 'Economic Scene', *New York Times*, 8 July.

Sprinzak, Ehud 1975, 'Marx's Historical Conception of Ideology and Science', *Politics and Society*, 5, no. 4: 395–416.

Suppe, Frederick (ed.) 1977, *The Structure of Scientific Theories* (2nd edn), Urbana: University of Illinois Press.

Thompson, E.P. 1978, *The Poverty of Theory and Other Essays*, New York: Monthly Review Press.

Thurow, Lester 1983, 'It's All Too Easy to Be a Critic', *New York Times*, 19 June.

Toulmin, Stephen 1982, 'The Construal of Reality: Criticism in Modern and Postmodern Science', *Critical Inquiry*, 9, no. 1: 93–111.

Wallace, Anthony F.C. 1980, 'Review of *Cultural Materialism*', *American Anthropologist*, 82, no. 2: 423–6.

Wallace, John 1979, 'Translation Theories and the Decipherment of Linear B', *Theory and Decision*, 11: 111–40.

Ward, Benjamin 1972, *What is Wrong With Economics?*, New York: Basic Books.

Weber, Max 1947, *The Methodology of the Social Sciences*, Glencoe: Free Press.

Weber, Max 1947, *The Theory of Social and Economic Organization*, New York: Free Press.

Williams, Raymond 1976, *Keywords*, New York: Oxford University Press.

Wilmsen, Edwin 1976, *Lindenmeier: A Pleistocene Hunting Society*, New York: Harper & Row.

Winch, Peter 1963, *The Idea of a Social Science and its Relation to Philosophy*, London: Routledge and Kegan Paul.

Winch, Peter 1977, 'Understanding a Primitive Society', in *Rationality*, edited by Bryan R. Wilson, Oxford: Blackwell.

Wittgenstein, Ludwig 1953, *Philosophical Investigations*, London: Macmillan.

Wolf, Eric 1966, 'Kinship, Friendship, and Patron-Client Relations in Complex Societies', in *The Social Anthropology of Complex Societies*, edited by Michael Banton, New York: Praeger.

Index

Printed in the United States
By Bookmasters